PROGRAMMED
BEGINNING ALGEBRA

UNIT VI Graphs and Linear Systems

AUTONETICS HOME STUDY PROGRAM

Organized and presented by Personnel Development with cooperation of the Technical Information Center.

Course: Beginning Algebra Volume 4

Course Number: B3 SZXHX

1. CHECKOUT TIME 4 WEEKS 30 HOURS
2. PREREQUISITES Volumes 1, 2, & 3 or equivalent
3. MATERIALS - ALL MATERIALS MUST BE RETURNED to the library except for your test answer sheet.

   Programmed Text
   Answer Sheet Booklet (contains special answer sheets for Vol. 4)
4. BE SURE TO READ AND FOLLOW ALL DIRECTIONS.
5. READ THE AUTHOR'S introduction and instructions.
6. PLEASE - DO NOT WRITE OR MAKE ANY MARKS ON THE TEXT.

   USE A SEPARATE SHEET of paper for responses and notes. Do write responses. It helps the learning process.
7. COMPLETE THE TEST and mail your answer sheet to Home Study, D/055, GB04
8. If you have comments, suggestions or need assistance, please call extension 2495

FOURTH PRINTING, JULY, 1967

Library of Congress Catalog Card Number: 63-11429

Printed in the United States of America

PREFACE

Programmed Beginning Algebra consists of eight units of material, covering topics normally a part of a course in first-year algebra. The material is organized as follows:

Volume I	Unit I	Natural Numbers
	Unit II	Integers
Volume II	Unit III	First-Degree Equations and Inequalities in One Variable
	Unit IV	Products and Factors
Volume III	Unit V	Fractions and Fractional Equations
Volume IV	Unit VI	Graphs and Linear Systems
Volume V	Unit VII	Radicals
	Unit VIII	Quadratic Equations

This program can be used in a variety of ways. For reasonably capable students who accomplish the entire program, it constitutes a complete first-year course in algebra. Preferably, the program should be supplemented with periodic classroom sessions or individual conferences with a qualified mathematics instructor; but, even without such assistance, a conscientious student can expect to achieve a degree of proficiency adequate to perform in a satisfactory manner on most standard achievement tests in beginning algebra.

Any one of the eight units can be used independently of the others. Each unit can be used as a self-tutor, to supplement and/or reinforce the same topic in any standard first-year algebra textbook. If classroom attendance is interrupted due to illness or for some other reason, the material missed in class can be covered by accomplishing the appropriate unit in this series. The statement of objectives and the table of contents at the beginning of each unit will give a good picture of the material covered herein.

The material, as presented, is somewhat modern in point of view. Set terminology and set symbolism are used where helpful, inequalities in one and two variables are discussed in conjunction with equations, solution sets of equations and their graphs are approached through the concept of ordered pairs, and emphasis is placed on the logical basis for the routine operations performed. Since the point of view of the program is modern, the material should also be quite helpful to persons who have previously studied algebra and who would like to review, and, at the same time, modernize their point of view toward the subject.

The units in the program are sequential, and, with the exception of Unit I, there are certain prerequisites necessary for each. Every volume, except Volume I, includes a prerequisite test, together with a suggested minimum score. This score is to be viewed as flexible. The primary purpose of the test is to provide the individual student with a means of determining whether or not he can expect to complete the material in the volume without undue difficulty.

IRVING DROOYAN
WILLIAM WOOTON

HOW TO USE THIS MATERIAL

The theory underlying programmed instruction is quite simple. The program presents you with information in small units called frames, and, at the same time, requires you to make frequent and active written responses in the form of a word, a phrase, or a symbol. Programmed material is only as effective as you, the student, make it. In order to achieve the maximum benefit from this unit, you should follow the following instructions carefully.

If you wish to measure what you learn, you should take one form of the self-evaluation tests at the end of Unit I before beginning the program. Then, after completing the unit, you can take the alternate form of the test and compare the two scores. The same procedure can be used for Unit II.

To progress through the program:

1. Place the response shield over the response column (the column on the left-hand side of the page) so that the entire column is covered.

2. Read the first frame on the right-hand side of the page carefully, noting, as you do, the place where you are asked to respond. Do not write a response until you have read the entire frame.

3. After deciding on the proper response, write it in the blank provided or at the end of the frame.

4. Slide the response shield down until it is level with the top of the next frame. This will uncover the correct response to the preceding frame and any additional remarks accompanying it.

5. Verify that you have made the correct response. You should almost always have done so, because the frames were constructed in such fashion that it will be difficult for you to make an error.

6. If you have made the correct response, repeat Steps 2 to 5 above with the next frame. If you have made an error on the frame, read the frame again, draw a line through your incorrect response and write the correct response. Following this, repeat Steps 2 to 5 above.

Essentially, this is all there is to your part of the task, and the program must assume the remainder of the responsibility for your learning. However, there are a few additional things you can do to improve the effectiveness of the material.

1. Do not try to go too fast. Each frame should be read carefully and you should think about the response you are going to make. Does it fit the wording of the frame? Does it make sense? In many frames you will find clues (called prompts) that are designed to aid you in making a correct response. Does your proposed response match the clue?

2. Mathematics is a language written in symbols. In order to understand mathematics, you have to make a conscious effort to read the symbolism as though it were written out in words.

3. Do not try to do too much at once. If you stay with your work too long at one time, you tend to become impatient, and the number of errors you make will probably rise. If you are making good progress, and are not tired, there is no reason to stop, but if you find yourself becoming bored, or find yourself making a high percentage of errors, take a break. Come back to the program later.

4. After taking a break, pick up your work a few frames behind where you stopped work. This will lead you back into the logical sequence with a minimum of confusion.

5. Do not be overly attentive to the remarks that accompany some of the responses in the response column. They are there to clear up points of confusion once in a while, but if you are not confused, don't waste time on them. The important thing as far as you are concerned is the response itself.

6. Complete the unit. If you do not complete the unit, you cannot expect to achieve the full benefit of the material.

UNIT VI Graphs and Linear Systems

OBJECTIVES

Upon completion of the unit, the student should:

1. Be able to define and identify examples of "axis," "coordinate," " "component," "ordered pair," "abscissa," "ordinate," "intercept," "graph," and "linear."

2. Be able to find solutions of first-degree equations in two variables.

3. Be able to graph first-degree equations in two variables.

4. Be able to solve simultaneous systems of first-degree equations in two variables by graphical methods.

5. Be able to solve simultaneous systems of first-degree equations by elimination.

6. Be able to identify inconsistent and dependent equations by inspection.

7. Be able to solve simple word problems expressible in terms of two first-degree equations in two variables.

8. Be able to graph the solution sets of simple linear inequalities in two variables.

9. Be able to read and write the set notation $\{(x,y) \mid \text{condition on } x \text{ and } y\}$.

CONTENTS

PREREQUISITE TEST

Complete this test and score your paper from the answers given at the end of this unit.

Given $x = 2$, $y = -3$, $z = 4$, find the numerical value for each expression.

1. $3xy - yz$
2. $\frac{x - 4z}{2}$
3. $\frac{3y - x - z}{3}$

Solve.

4. $3(x - 4) + 1 = 10$
5. $\frac{x - 2}{3} = x - 6$
6. $\frac{3}{4}(x + 1) = \frac{1}{3}x + 2$

Solve for y in terms of x.

7. $2x = 4 - y$
8. $2x + 3y = 4$
9. The ratios $\frac{2}{3}$ and $\frac{18}{24}$ (are/are not) in proportion.
10. The ratios $\frac{4}{7}$ and $\frac{24}{42}$ (are/are not) in proportion.

If you missed more than three problems on this test, you are probably not adequately prepared to start Unit VI. We suggest you prepare yourself by working through earlier units of Programmed Beginning Algebra.

UNIT VI
GRAPHS AND LINEAR SYSTEMS

Remark. This unit is concerned with relationships that exist between two sets of numbers. These relationships are expressed by means of equations in two variables, or visually in terms of graphs. You should be familiar with some such relationships. For example, you probably already know that the circumference of a circle is just π times the length of the diameter, and you have probably used the formula $C = \pi D$, which states this relationship.

The first few frames in this unit will introduce some vocabulary that we will be using as we go along.

equation

1. Recall that when two algebraic expressions are connected by an "equals" sign, the result is called an equation. $3x - 3 = 2x$ is an ______________.

variables

2. Variables are symbols such as x, y, z that represent any member of a given set of numbers. In the equation $y = x + 3$, the symbols y and x are________________.

constant

3. Constants are symbols such as 3, 5, and 20 that represent but one number during a discussion. In the equation $y = x + 3$, the symbol 3 is a______________.

variables; constant

4. In the equation $y = x + 8$, x and y are ________________ and 8 is a ______________.

ANSWERS TO TESTS

Prerequisite Test

1. −6	2. −7	3. −5	4. 7	5. 8
6. 3	7. $y = 4 - 2x$	8. $y = \frac{4 - 2x}{3}$	9. are not	10. are

v; t; 32

5. In the equation $v = 32t$, _____ and _____ are variables and _____ is a constant.

first-degree

6. $3x - 3 = 2x$ is a first-degree equation in one variable because only one variable appears, and that variable has an exponent of 1. $x + 7 = 2x$ is also a ____________ equation in one variable.

equation; one

7. $x + 2 = 8$ is a first-degree _________ in _____ variable.

two

8. $y = x - 8$ is a first-degree equation in two variables because two variables are present and each variable has an exponent of 1. $y = 7 - x$ is also a first-degree equation in _____ variables.

two

9. $x + y = 4$ is a first-degree equation in _____ variables.

one

10. $x + 6 = 8$ is a first-degree equation in _____ variable(s).

64

11. An equation in two variables serves to pair numbers. In the equation $y = 32x$, an assignment of 2 for x gives $y = 32(2)$. For the left member to be equal to the right member, requires that the value of y be _____.

96

12. An assignment of 3 for x in the equation $y = 32x$ results in $y = 32(3) = 96$. Therefore, the value of 3 for x is paired with the value _____ for y.

128

13. The equation $y = 32x$ pairs the value 4 for x with the value _____ for y.

FINAL EXAM, UNIT VI

Questions 1 and 2 refer to Figure 1.

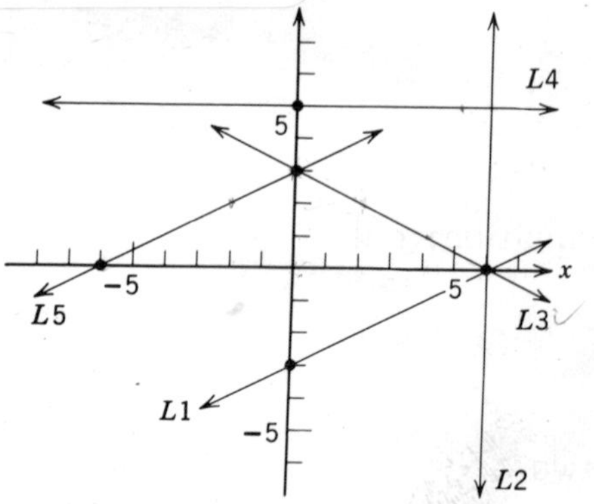

1. The graph of $x + 2y = 6$ is L-_____.

2. The graph of $x - 2y = -6$ is L-_____.

3. The solution of the system $\begin{matrix} x - y = 6 \\ x + y = 4 \end{matrix}$ is (___, ___).

4. The solution of the system $\begin{matrix} 3x - y = 11 \\ x + 2y = 6 \end{matrix}$ is (___, ___).

5. The solution of the system $\begin{matrix} y = 2x + 3 \\ 3x = y - 4 \end{matrix}$ is (___, ___).

6. The equations of the system $\begin{matrix} 3x - 2y = 6 \\ 9x - 6y = 18 \end{matrix}$ are (inconsistent/dependent).

7. The price of a dress and a sweater together totals \$22. If the price of the dress is \$4 greater than twice the price of the sweater, the price of the dress is________ and the price of the sweater is________.

8. Graph $2x + y = 4$ on the axes provided.

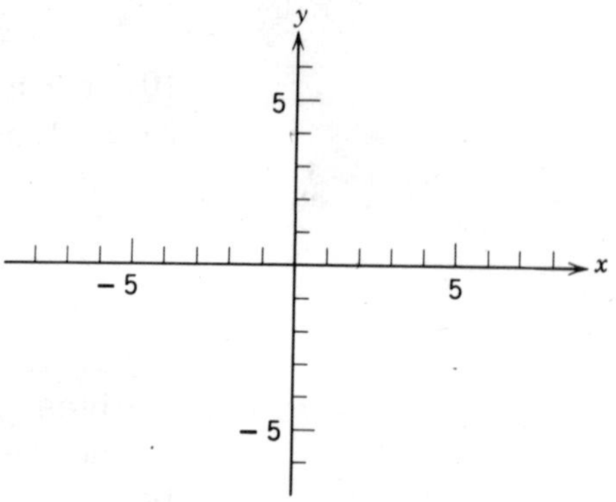

9. Graph $y > x - 3$ on the axes provided.

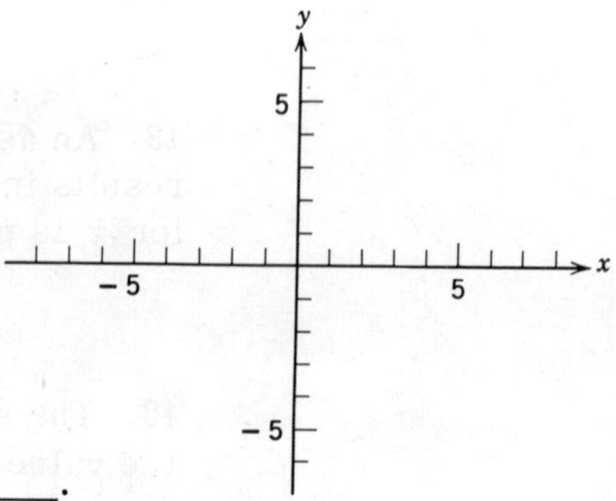

10. The solution set of $y > x - 3$ appears in set notation as____________________.

3

14. The equation $y = x + 2$ pairs the value 1 for x with the value _____ for y because $y = (1) + 2$.

−4

15. To find numbers that are paired by an equation in two variables, any value can be assigned to one of the variables, and the value of the other variable can then be determined. For example, if x is assigned the value 0, the equation $y = x - 4$ pairs the number 0 with the number ______.

10

16. If x is assigned the value 5, the equation $y = 2x$ pairs with 5 the number ________.

6

17. If the value −2 is assigned to x, the equation $x + y = 4$ becomes $-2 + y = 4$. The solution of this last equation is 6, so that $x + y = 4$ pairs the numbers −2 and ______.

7

18. If x is assigned the value −3, the equation $x + y = 4$ pairs with −3 the number_____.

ordered

19. To simplify writing pairs of numbers that are values for two different variables, the numbers are written in parentheses, separated by a comma, and called ordered pairs. Thus, (3,2) is an ordered pair and (5,2) is an________________pair.

ordered pair

20. (2,−3) is an______________ ___________.

3

21. In any ordered pair, the number on the left is called the first component and the number on the right is called the second component. In the ordered pair (2,3), 2 is the first component and _____ is the second component.

This concludes Unit VI and Volume 4

COMPLETE THE TEST USING ONE OF THE ANSWER SHEETS PROVIDED--YOU CAN'T LOSE

a. AFTER WE RECEIVE YOUR COMPLETED TEST, A RECORD OF SUCCESSFUL completion will go to your supervisor and into your personnel folder. Also, you will receive a copy. These records will reflect your personal effort and will inform management of new capabilities.

b. NO RECORD will be kept if your test result is unsatisfactory. After all, you are taking the course voluntarily and on your own time. You will be contacted after your test is scored if it is unsatisfactory.

<u>Please mail your completed test answer sheet to:</u>

Home Study   D/055   031-GB04

When you return this text to the library, you may wish to check out Volume 5, the next part of this sequential course.

first component

22. In the ordered pair $(-1,-7)$, -1 is the __________ ______________ and -7 is the second component.

-7

23. The first component of the ordered pair $(3,-7)$ is 3 and the second component is_____.

-5; 7

24. In the ordered pair $(-5,7)$, the first component is______ and the second component is_____.

is not

25. The name "ordered pair" comes from the fact that the order in which the components are written is important. Thus, the ordered pair $(3, 2)$ is not the same as the ordered pair $(2, 3)$ because the pairs do not have the same first and second components. The ordered pair $(1, 5)$ (is/is not) the same as $(5, 1)$.

is not

26. The ordered pair $(2,2)$ (is/is not) the same as $(3, 3)$.

2

27. If the components of an ordered pair represent replacements for the variables x and y, it is customary to write the value of x first in the ordered pair, and the value of y second. For example, the ordered pair $(3, 2)$ means that the value of x is 3 and the value of y is______.

y-component

28. In the ordered pair $(2,-3)$, the x-component is 2 and the y-____________________ is -3.

x; y

Or first; second.

29. In the ordered pair $(0, 4)$, the_____-component is 0 and the_____-component is 4.

$y = 5 - x$; $y > 5 - x$

561. The inequality $y \geq 5 - x$ includes in its solution set the solution sets of both $y = 5 - x$ and $y > 5 - x$. Thus, the graph of $y \geq 5 - x$ will consist of the graphs of both $y =$ __________ and $y >$ __________ and is shown as

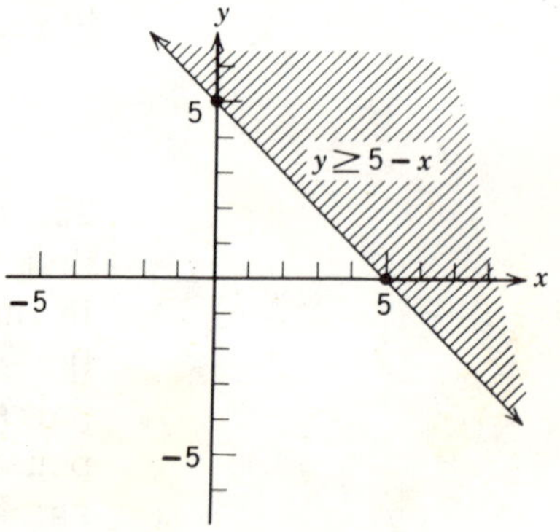

where the graph of $y = 5 - x$ is shown as a solid line.

$y > x - 2$

562. The notation $\{(x, y) \mid y = x - 2\}$, read "the set of ordered pairs (x, y) such that $y = x - 2$," is used to describe the infinite set of solutions of the equation $y = x - 2$. Similarly, $\{(x, y) \mid y > x - 2\}$ is used to describe the solution set of the inequality

__________.

$\{(x, y) \mid y \leq 3x + 2\}$

563. The solution set of $y \leq 3x + 2$ can be shown in set notation as $\{$ __________ $\}$.

Remark. This concludes Unit VI. To see what you have learned, you can now accomplish one form of the evaluation test that follows. If you completed one form of this test prior to beginning the unit, use the other form now, and compare the two scores.

(6, −5)

30. Write an ordered pair whose x-component is 6 and whose y-component is -5.

solution

31. In an equation in two variables, if the variables are replaced with the components of an ordered pair and the resulting statement is true, the ordered pair is called a solution of the equation. If the components of (2, 1) are substituted in $x + 2y = 4$, the result is $(2) + 2(1) = 4$ or $4 = 4$. Therefore, (2,1) is a ________________ of $x + 2y = 4$.

solution

32. Replacing the variables x and y in the equation $y = 2x + 4$ by the respective components of the ordered pair (3, 10) results in $10 = 2(3) + 4$ or $10 = 10$, which is true. The ordered pair (3, 10) is, therefore, a ________________ of the equation.

true

33. (1, 6) is a solution of $y = 2x + 4$ because replacing the variables by their respective components results in $6 = 2(1) + 4$ or $6 = 6$, which is (true/false).

solution

34. Substituting -3 for x and -2 for y in the equation $y = 2x + 4$ gives $-2 = 2(-3) + 4$ or $-2 = -2$. Therefore, (−3, −2) is a ________________ of the equation.

is

If x is replaced by -1 and y by 2, the result is $2 = 2(-1) + 4$ or $2 = 2$.

35. (−1, 2) (is/is not) a solution of $y = 2x + 4$.

is not

Because $1 = 2(-2) + 4$ or $1 = 0$ is false.

36. (−2, 1) (is/is not) a solution of $y = 2x + 4$.

are

555. The equations in the system

$$3x + y = 5$$
$$6x - 2y = 8$$

(are/are not) inconsistent.

is

556. Inequalities in two variables such as $y > x + 3$ have an infinite number of solutions. (3, 7) is a solution because when x is replaced by 3 and y by 7 the resulting statement $7 > 3 + 3$ is true. (3, 8) (is/is not) a solution of $y > x + 3$.

6

557. If x represents 3, the inequality $y > x + 3$ can be written $y > 3 + 3$ or $y > 6$, which is true as long as y is greater than_____.

17

558. The ordered pair $(8, y)$ will be a solution of $y > 2x + 1$ as long as y is greater than______.

5

559. The ordered pair $(0, y)$ will be a solution of $y < 5 - x$ as long as y is less than_____.

below; above

560. The graph of $y = 5 - x$ consists of all points on a line. The graph of $y < 5 - x$ consists of all points in the region on the plane__________the graph of $y = 5 - x$. The graph of $y > 5 - x$ consists of all points in the region on the plane__________the graph of $y = 5 - x$.

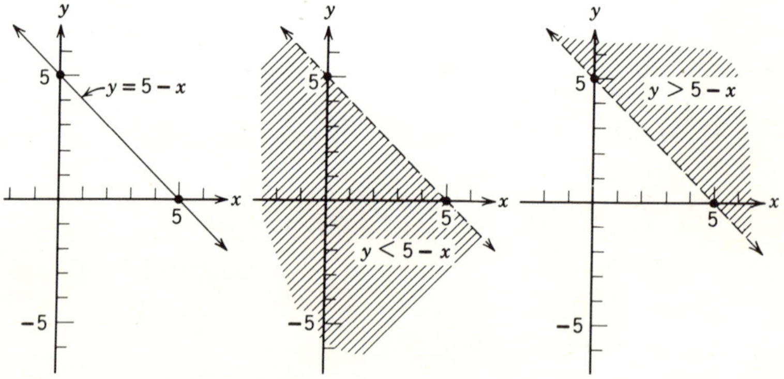

is

$14 = 2(5) + 4$
$14 = 14.$

37. (5, 14) (is/is not) a solution of $y = 2x + 4$.

x; y

38. (2, 1) is a solution of $2x + y = 5$; 2 is a replacement for the variable___, and 1 is a replacement for the variable____.

is

39. The ordered pair (0, 8) (is/is not) a solution of $2x + y = 8$.

is

40. The ordered pair (4, 0) (is/is not) a solution of $2x + y = 8$.

15

Because if x is replaced with 5, $y = 3(5)$ or $y = 15$.

41. If the ordered pair $(3, y)$ is to be a solution of $y = 3x$, then the value of y can be found by substituting 3 for x in the equation. Thus, y must have the value 9. If the ordered pair $(5, y)$ is to be a solution of $y = 3x$, then y must have the value______.

(−1, −3)

Because if x is replaced with −1, $y = 3(-1)$ or $y = -3$.

42. A solution of $y = 3x$ is (−1, ____).

(0, 0)

Because $y = 3(0)$ or $y = 0$.

43. A solution of $y = 3x$ is (0, ____).

(−1, 4)

44. A solution of $y = 3x + 7$ is (−1, ____).

(7, 2)

551. The system

$$x - 3y = 1$$
$$3x + \quad y = 23$$

can be transformed to the equivalent system

$$-3x + 9y = -3$$
$$3x + \quad y = 23$$

by multiplying each member of the first equation by −3. The solution of both systems is (___, ___).

are

552. In a system of two equations in two variables, if the coefficient of the corresponding terms in each equation are proportional, then the equations are either inconsistent or else dependent. The equations in the system

$$3x + 2y = a$$
$$6x + 4y = b$$

(are/are not) either inconsistent or dependent.

inconsistent

553. The equations in the system

$$2x + 3y = a$$
$$4x + 6y = b$$

are dependent if $\frac{a}{b} = \frac{2}{4} = \frac{3}{6}$ and they are inconsistent if $\frac{a}{b} \neq \frac{2}{4} = \frac{3}{6}$. If $a = 6$ and $b = 9$, the equations are

________________.

dependent

554. The equations in the system

$$x - 3y = 4$$
$$3x - 9y = 12$$

are (inconsistent/dependent).

(3, 16)	45. A solution of $y = 3x + 7$ is (3, _____).
(0, 7)	46. A solution of $y = 3x + 7$ is (0, _____).
(−5, −8)	47. A solution of $y = 3x + 7$ is (−5, _____).
infinite	48. Since an equation in two variables pairs a number with every value assigned to one variable, an equation such as $y = 3x + 7$ has an infinite number of solutions. Similarly, $y = x + 2$ has an ________ number of solutions.
infinite	49. $y = x - 5$ has an __________ number of solutions.
(−3, −8)	50. A solution of $y = x - 5$ is (−3, _____).
(−2, −7)	51. A solution of $y = x - 5$ is (−2, _____).
(0, −5)	52. A solution of $y = x - 5$ is (0, _____).
(1, −4)	53. A solution of $y = x - 5$ is (1, _____).
(3, −2)	54. Some solutions of $y = x - 5$ are (−3, −8), (−2, −7), (0, −5), and (1, −4). Another solution is (3, _____).
(1, 2); (2, 4); (3, 6)	55. Some solutions of $y = 2x$ are (−3, −6), (−2, −4), (−1, −2), (0, 0), (1, _____), (2, _____), and (3, _____).

$2x = 8$

548. The addition axiom of equality asserts that if

$$a = b$$

and

$$c = d$$

are true, then

$$a + c = b + d.$$

Thus, for any values of the variables for which

$$x + y = 2$$
$$x - y = 6$$

are both true, the addition axiom asserts that

_____ = _____.

(4, −2)

549. The addition axiom of equality applied to the system

$$x + y = 2$$
$$x - y = 6$$

gives $2x = 8$. This equation is then used to find one component of the solution of the system. The complete solution of the system is the ordered pair

__________.

(3, 2)

550. The system

$$2x + y = 8$$
$$x + y = 5$$

can be transformed to the equivalent system

$$2x + y = 8$$
$$-x - y = -5$$

by multiplying each member of the second equation by −1. The solution of both systems is (___, ___).

$(1,2)$; $(2,5)$; $(3,8)$

56. Some solutions of $y = 3x - 1$ are $(-3,-10)$, $(-2,-7)$, $(-1,-4)$, $(0,-1)$, $(1, \_\_\_\_)$, $(2, \_\_\_\_)$, and $(3, \_\_\_\_)$.

$(5,3)$; 3

57. The equation $y = 3$ may be visualized as an equation in two variables, $y = 0x + 3$. Some solutions of this equation are $(1,3)$, $(2,3)$, $(3,3)$, and $(4,3)$. Another solution is $(5, \_\_\_\_)$. The second component (replacement for y) is always_____.

-4

58. The equation $y = -4$ may be visualized as $y = 0x - 4$. Some solutions of this equation are $(0,-4)$, $(2,-4)$, and $(6,-4)$. The second component is always_____, regardless of the value of the first component.

$(5,4)$; 5

59. The equation $x = 5$ may be visualized as an equation in two variables, $0y + x = 5$. Some solutions of this equation are $(5,1)$, $(5,2)$, and $(5,3)$. Another solution is $(\_\_\_\_, 4)$. The first component (replacement for x) is always_____.

-2

60. The equation $x = -2$ may be visualized as $0y + x = -2$. Some solutions of this equation are $(-2,0)$, $(-2,5)$, and $(-2,10)$. The first component is always_____, regardless of the value of the second component.

point

542. Any point in the plane can be associated with a unique ordered pair of numbers. Since a solution of an equation in two variables is an ordered pair of numbers, such a solution can be associated with a single__________in the plane, called the graph of the ordered pair of numbers.

infinite

543. An equation in two variables has an infinite number of solutions (ordered pairs of numbers). Therefore, the members of the solution set are associated with an ____________number of points that are called the graphs of the solutions.

graph

544. The infinite number of points associated with the solutions of a first-degree equation in two variables lie on a straight line, and the line is said to be the__________of the equation.

(0, −8); (4, 0)

545. Two points are necessary to determine a straight line. Using two solutions in which at least one of the components is zero is called the intercept method of graphing. Two solutions of $y = 2x - 8$ in which at least one component is zero are (0, ______) and (_____, 0).

(0, −3); (5, 0)

546. Two solutions of $3x - 5y = 15$ in which at least one of the components is zero are__________and __________.

intersection

547. Two equations considered together are called a system of simultaneous equations, and if the system has a single solution, the solution is the ordered pair of numbers associated with the coordinates of the point of__________________of their graphs.

Remark. If it helps make things clear to you, equations in two variables can be thought of as "machines" whose job it is to manufacture ordered pairs. If you feed a value of x into the machine, it pairs the value of x with a value for y and produces an ordered pair (x, y). If the equation is of the first degree, it will produce a value of y for every value of x. Since each ordered pair it produces is a solution of the equation, the solution set of the equation contains an infinite number of members.

For the moment, however, we are going to be more concerned with specific solutions of equations than we are with the entire solution set. We shall want to be able to find one or two particular solutions whenever the necessity arises. One of the techniques we use for this purpose involves equivalent equations.

equivalent

61. You should recall from Unit III that equivalent equations are any equations having the same solution set, and that adding or subtracting the same quantity to or from each member of an equation, or multiplying or dividing each member of an equation by the same nonzero quantity, will produce an equation equivalent to the one with which you started. Thus, $x + 5 = 7$ and $x = 2$ are__________________equations because they have the same solution set.

solution; adding

62. $x - 2 = 5$ and $x = 7$ are equivalent equations because they have the same_______________ set. $x - 2 = 5$ can be transformed to $x = 7$ by _______________2 to each member of $x - 2 = 5$.

equivalent equations; multiplying

63. $\frac{x}{3} = 5$ and $x = 15$ are__________ __________.
The first equation can be transformed to the second equation by__________________each member of $\frac{x}{3} = 5$ by 3.

dividing

64. $-3y = 18$ and $y = -6$ are equivalent equations. The first equation can be transformed to the second equation by______________ each member of $-3y = 18$ by -3.

-2	**536.** An equation in two variables serves to pair numbers. For example, the equation $y = x + 2$ will pair the number 7 with the number 5. If x is assigned the value -4, then the equation will pair this number with the value _____ for y.
solution	**537.** If each of the variables of an equation in two variables is replaced with a number, and if the resulting statement is true, then the pair of numbers constitutes a solution of the equation. Since the replacement of x with 5 and y with 7 in the equation $y = x + 2$ results in the true statement $7 = 5 + 2$, the numbers 5 and 7, considered together, constitute a ____________ of the equation.
ordered pair	**538.** One way of displaying a solution of an equation in two variables is to list the numbers in a solution as an ordered pair. (5, 7) is an example of an __________ _________.
x; y	**539.** In any equation in two variables in which the variables x and y are used, solutions are customarily given by means of ordered pairs of the form (x, y). For example, (3, 2) is a solution of $x + 2y = 7$, where the first component, 3, is a replacement for_____ and the second component, 2, is a replacement for______.
5; (2, 5)	**540.** Any number of solutions of an equation such as $y = 3x - 1$ can be found by assigning a value to x, and computing an associated value for y. For example, if x is assigned the value 2, then the associated value for y is_____, and (2, _____) is a solution of the equation.
ordered pair	**541.** The set of all solutions of an equation in two variables is called the solution set of the equation. Each member of the set is an __________ ______ of numbers.

equivalent; adding

65. $y - x = 5$ and $y = x + 5$ are______________ equations because they have the same solution set. The second equation can be obtained from the first by____________ x to each member.

$-2x$

66. $y + 2x = 8$ can be written as the equivalent equation $y = 8 - 2x$ (where the left member consists entirely of y) by adding________to each member.

2

67. $2y = x + 5$ can be rewritten as the equivalent equation $y = \frac{x + 5}{2}$ by dividing each member of $2y = x + 5$ by_____.

solved

68. When an equation is rewritten with a variable having a coefficient of one as the only term in one member, the equation is said to be solved for that variable. When $2y = x + 5$ has been rewritten as $y = \frac{x + 5}{2}$, the equation has been____________for y.

$y = 6 + 3x$

$$y - 3x = 6$$
$$y - 3x + 3x = 6 + 3x$$
$$y = 6 + 3x.$$

69. Rewrite $y - 3x = 6$ as an equivalent equation in which the left member consists of y alone. That is, solve for y.

$y = \frac{8 - 3x}{2}$

$$3x + 2y = 8$$
$$2y = 8 - 3x$$
$$y = \frac{8 - 3x}{2}.$$

70. Solve $3x + 2y = 8$ for y.

true

71. One solution of $y = \frac{8 - 3x}{2}$ is $(0, 4)$, because replacing x by 0 and y by 4 results in $4 = \frac{8 - 3(0)}{2}$ or $4 = 4$, which is (true/false).

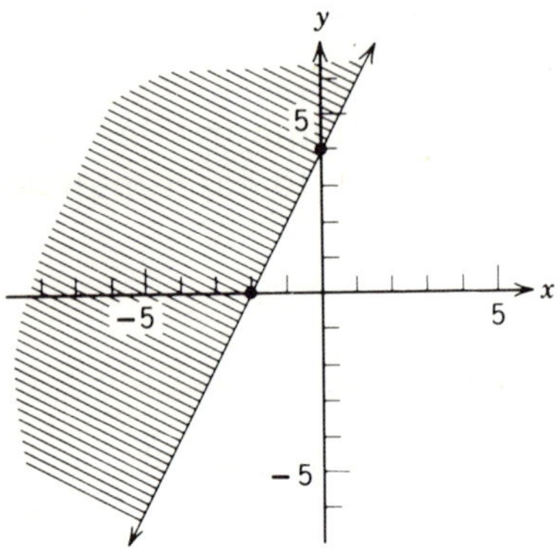

534. Graph $\{(x, y) \mid y \geq 2x + 4\}$.

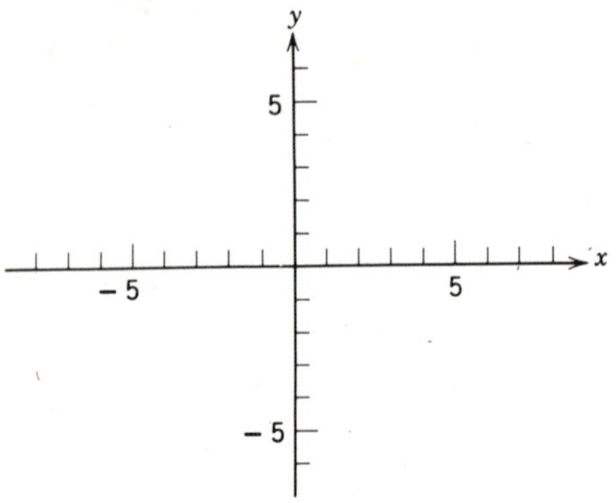

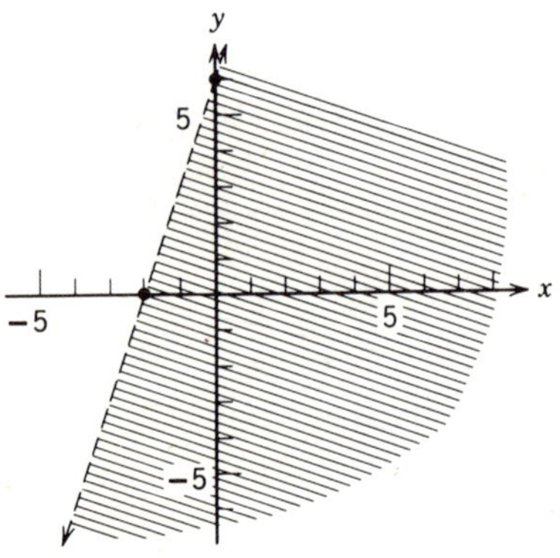

535. Graph $\{(x, y) \mid y < 3x + 6\}$.

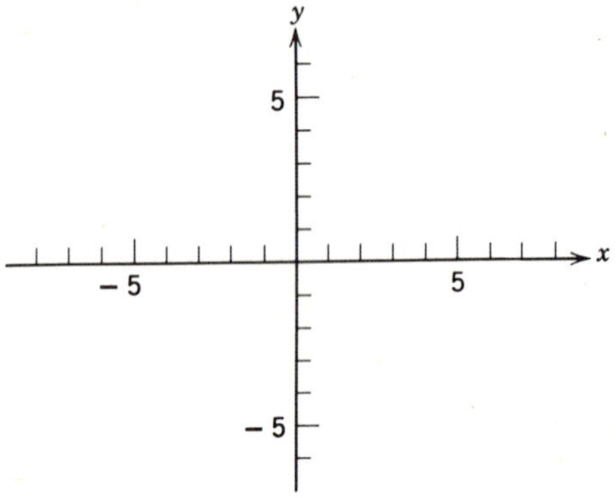

Remark. Graphs are sometimes used to find solution sets of systems of simultaneous inequalities, the solution sets of such a system being determined by the points that lie in the graphs of every inequality in the system. Perhaps you will find yourself working with such systems when you study more advanced mathematics.

We are going to turn our attention now to a review. The following sequence of frames will give you an opportunity to obtain an overview of this unit in a very short time. Although you should not anticipate learning anything you have not learned to this point, the review will highlight the important ideas in a brief way and will reinforce what you have learned. If you feel the need of a review of the solution of word problems, we suggest you quickly reread Frames 404 to 449.

(2, 1)

72. The replacement of x by 2 in the equation $y = \frac{8 - 3x}{2}$ results in $y = \frac{8 - 3(2)}{2}$ or $y = 1$. Therefore, (2, ____) is a solution.

(4, −2)

$y = \frac{8 - 3(4)}{2}$

$y = -2.$

73. (4, ____) is a solution of $y = \frac{8 - 3x}{2}$.

(6, −5); (8, −8)

74. Some solutions of $y = \frac{8 - 3x}{2}$ are (0, 4), (2, 1), and (4, −2). Additional solutions are (6, ____) and (8, ____).

$y = \frac{2 + x}{3}$

75. Solve $3y - x = 2$ for y.

(10, 4)

76. One solution of $y = \frac{2 + x}{3}$ is (4, 2). Another solution is (10, ____).

$y = \frac{x - 8}{2}$

77. Solve $2y - x + 8 = 0$ for y.

$\left(1, \frac{-7}{2}\right)$

78. One solution of $y = \frac{x - 8}{2}$ is (0, −4). Another solution is $\left(1, \_\_\_\_\right)$.

(−2, −5); (0, −4)

79. Given the equation $y = \frac{x - 8}{2}$, find the second components of each of the solutions (−2, ____) and (0, ____).

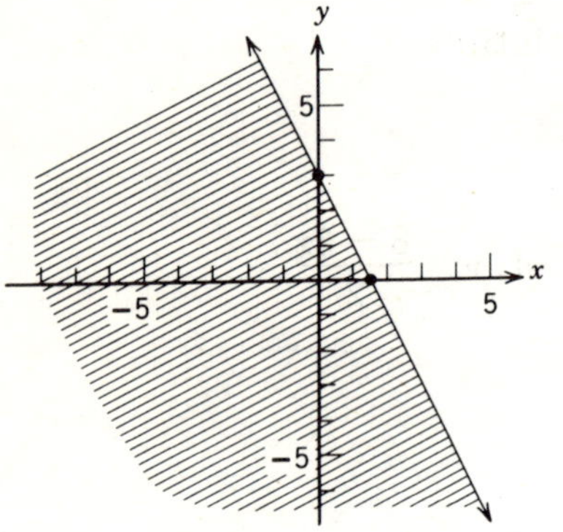

531. The figure below shows the graph of $\{(x, y) \mid y = 3 - 2x\}$. Graph $\{(x, y) \mid y \leq 3 - 2x\}$ on the same set of axes.

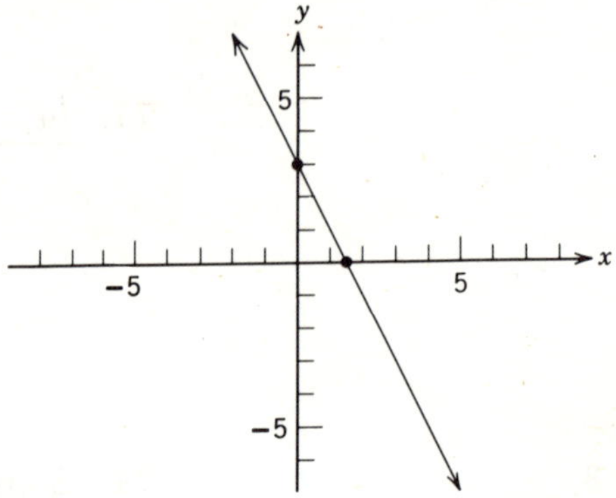

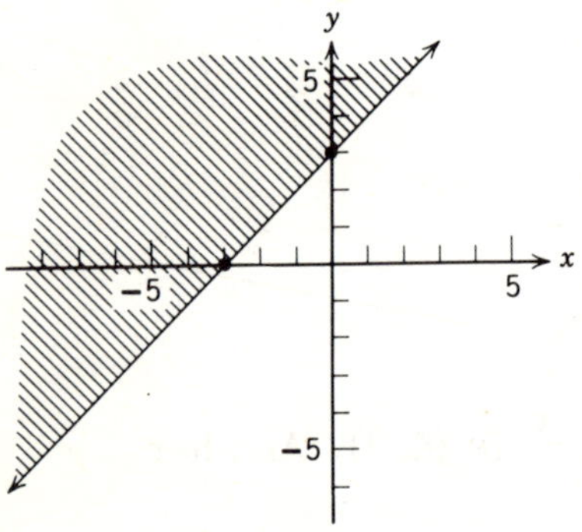

532. Graph $\{(x, y) \mid y \geq x + 3\}$.

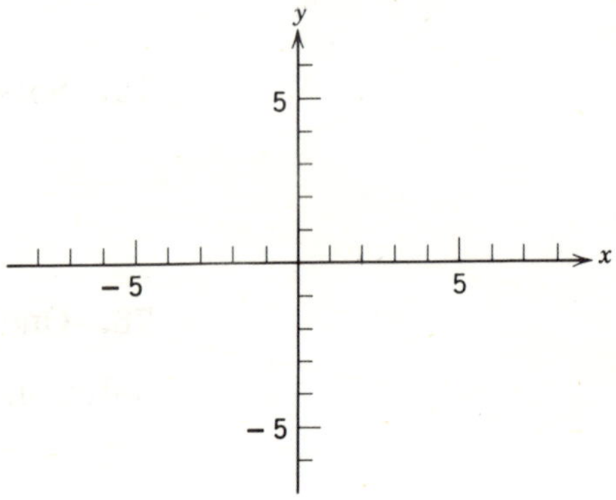

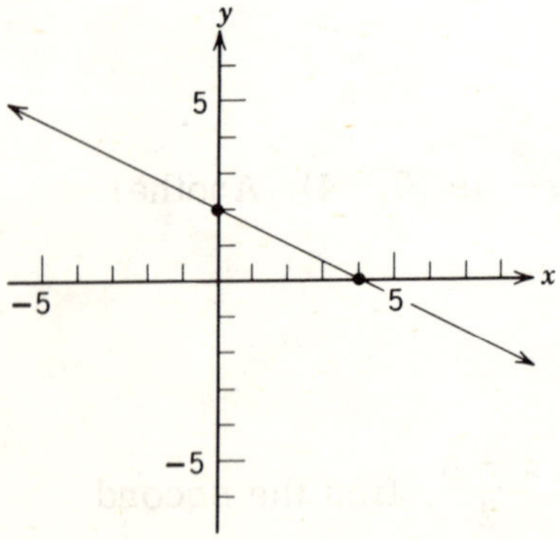

533. Graph $\{(x, y) \mid x + 2y = 4\}$.

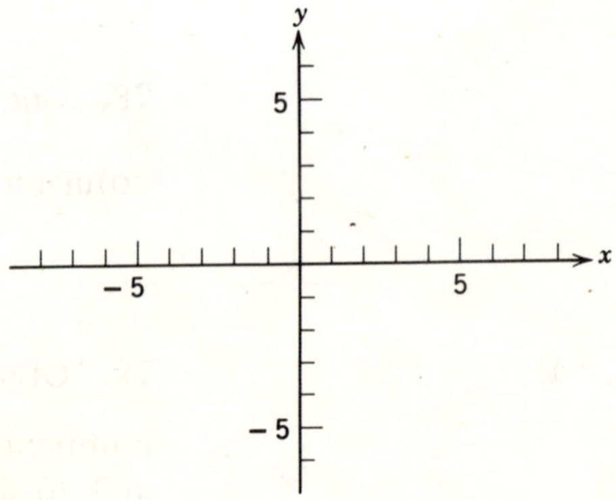

$-2x$

80. The equation $2x - y = 4$ can be rewritten as an equivalent equation $-y = 4 - 2x$ by adding ________ to each member of $2x - y = 4$.

$y = -4 + 2x$

Or $y = 2x - 4$.

81. Multiplying each member of $-y = 4 - 2x$ by -1 yields $y =$ ____________.

$(0, -4)$; $(5, 6)$; $(10, 16)$

82. Given the equation $y = 2x - 4$. Find the second components of each of the solutions (0, ______), (5, ______), and (10, ______).

$(0, 4)$; $\left(2, \frac{8}{3}\right)$; $(6, 0)$

Did you first solve for y?

83. Given the equation $2x + 3y = 12$. Find the second components of each of the solutions (0, ______), (2, ______), and (6, ______).

$(7, 2)$

84. Given that the second component of a solution of the equation $x + y = 9$ is 2, the first component can be determined by replacing y by 2 and solving for x. The desired solution is (______, 2).

$(5, 4)$

85. A solution of $x + y = 9$ is (______, 4).

$(9, 0)$

86. A solution of $x + y = 9$ is (______, 0).

$(0, 2)$

87. A solution of $x + 3y = 6$ is (0, ______).

$(0, 4)$

88. A solution of $2x + 3y = 12$ is (0, ______).

$\{(x, y) \mid y = 3x - 1\}$

524. The solution set of $y = 3x - 1$, which includes an infinite number of ordered pairs of numbers, is denoted by $\{$ ______________________ $\}$.

ordered pairs

525. The solution set of $x + 2y < 3$ is denoted by $\{(x, y) \mid x + 2y < 3\}$. This symbolism represents an infinite number of ____________ ________ of numbers.

$y \geq 3 - x$

526. $\{(x, y) \mid y \geq 3 - x\}$ denotes the solution set of ____________.

$y \leq 4 - 2x$

527. $\{(x, y) \mid y \leq 4 - 2x\}$ denotes the solution set of ____________.

$\{(x, y) \mid y \leq 3x + 1\}$

528. Use set notation to write the solution set of $y \leq 3x + 1$.

$\{(x, y) \mid y = 3 - 2x\}$

529. Use set notation to write the solution set of $y = 3 - 2x$.

$\{(x, y) \mid x + y \leq 3\}$

530. Use set notation to write the solution set of $x + y \leq 3$.

Remark. Because the solution sets of equalities and inequalities can be displayed both by graph and by use of set notation, the set notation offers another means of determining a graph. The next few frames of this unit will show how this goes.

Remark. It is important that you be able to find solutions of equations in two variables, because this ability will be put to good use in what follows. We shall next consider the association that exists between ordered pairs of numbers and the points in a geometric plane, an association that provides us with the most concrete means imaginable for visualizing mathematical relationships.

real

89. Every point on a line can be associated with a number. The infinite set of numbers associated with the infinite number of points on a line is called the set of real numbers. Thus, −5, −3/2, 0, 4, and 7.3 may be associated with points on a line and are ________numbers. (The word "real" is used in a technical sense, and should not be interpreted as meaning that there are "unreal" numbers.)

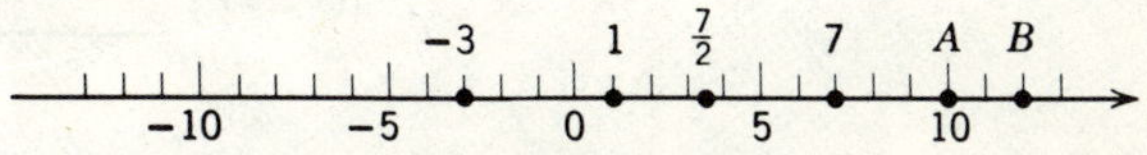

Fig. 1

real numbers

90. [Figure 1] The numbers −3, 1, 7/2, and 7 are associated with points on a line. These numbers are ________ ________.

coordinate

91. [Figure 1] The points on the line are called the graphs of the numbers and the numbers are called the coordinates of the points. 10 is the ________ of the point marked *A*.

12

92. [Figure 1] The coordinate of the point marked *B* is________.

$x - 3y = 4$

516. The solution set of the equation $x - 3y = 4$ may be written $\{(x, y) \mid$ ______________ $\}$.

$\{(x, y) \mid 3y + 2x = 4\}$

517. The solution set of the equation $3y + 2x = 4$ may be written ____________________.

$s = 3r + 2$

518. A variety of symbols may be used as variables. $\{(r, s) \mid s = 3r + 2\}$ represents the solution set of the equation ______________.

$a = 3b - 1$

519. The solution set of $a = 3b - 1$ may be written $\{(a, b) \mid$ ______________ $\}$.

$y > 2x + 3$

520. The set notation $\{(x, y) \mid y > x + 1\}$ is read "the set of ordered pairs (x, y) such that y is greater than $x + 1$," and denotes the solution set of the inequality $y > x + 1$. Using the same notation, the solution set of $y > 2x + 3$ is shown by $\{(x, y) \mid$ ______________ $\}$.

greater than

521. $\{(x, y) \mid x + y > 1\}$ is read "the set of ordered pairs (x, y) such that x plus y is __________ __________ 1."

such that

522. $\{(x, y) \mid x - y \leq 0\}$ is read "the set of ordered pairs (x, y) __________ __________ x minus y is less than or equal to 0."

(x, y)

523. The solution set of $y < 2x + 3$ is denoted by $\{($ ___ , ___ $) \mid y < 2x + 3\}$.

−6

93. [Figure 2] A line graph can be constructed in a vertical position as illustrated. The coordinate of the point marked *A* is ________.

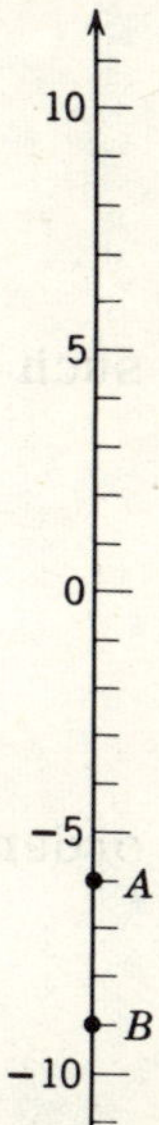

Fig. 2

coordinate

94. [Figure 2] −9 is the ________ of the point marked *B*.

origin

95. The point corresponding to 0 on a line graph is called the origin. Thus, in either of the line graphs [Figure 1] or [Figure 2], the symbol 0 indicates the ________.

right

96. In constructing a horizontal line graph, an arrowhead is used on the right-hand side to indicate that this is the direction of increasing value associated with the points. Thus, of any two real numbers, the point associated with the larger number is located to the (right/left) of the point associated with the smaller number.

above

97. In constructing a vertical line graph, an arrowhead is used on the upper part of the line to indicate that this is the direction of increasing values associated with the points. Thus, of any two real numbers, the point associated with the larger number is located (above/below) the point associated with the smaller number.

below

−5 is less than 3.

98. On a vertical line graph, the point corresponding to −5 will be located (above/below) the point corresponding to 3.

set

508. A special notation, $\{(x, y) \mid y = x + 2\}$, is used to describe the infinite number of ordered pairs making up the solution set of the equation $y = x + 2$. The braces refer to a set of any kind; thus, $\{(x, y) \qquad \}$ refers to a ________ of ordered pairs.

such that

509. In the notation $\{(x, y) \mid y = x + 2\}$, the vertical bar "$|$" is read "such that." Thus, the complete symbolism is read "the set of ordered pairs (x, y) ________ ________ y equals x plus 2."

ordered

510. $\{(x, y) \mid y = x + 2\}$ is read "the set of ________ pairs (x, y) such that y equals x plus 2."

set

511. $\{(x, y) \mid x - y = 3\}$ is read "the ________ of ordered pairs (x, y) such that x minus y equals 3."

solution

512. $\{(x, y) \mid y = 2x\}$ is a compact notation describing an infinite number of ordered pairs making up the solution set of the equation $y = 2x$. One of the ordered pairs is $(3, 6)$ because it is a ________ of the equation $y = 2x$.

$y = 2x$

513. $\{(x, y) \mid y = 2x\}$ denotes the solution set of the equation ________.

solution set

Braces, { }, always denote a set.

514. $\{(x,y) \mid x + y = 5\}$ denotes the ________ ________ of the equation $x + y = 5$.

solution set; $2x + 3y = 0$

515. $\{(x,y) \;\; 2x + 3y = 0\}$ denotes the ________ ________ of the equation ________.

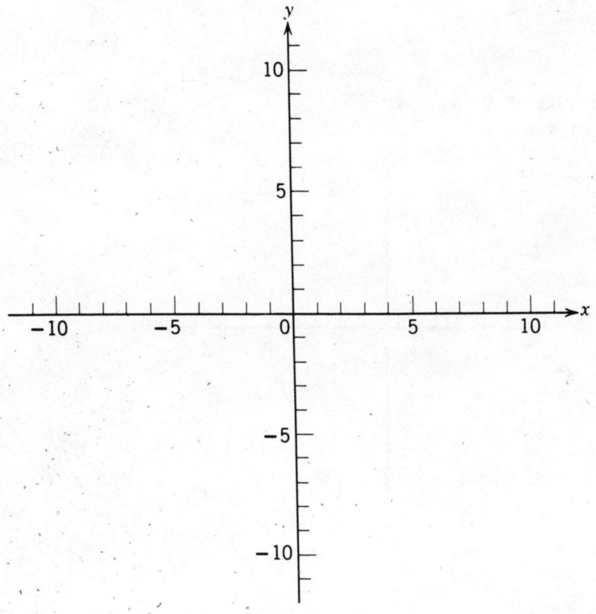

Fig. 3

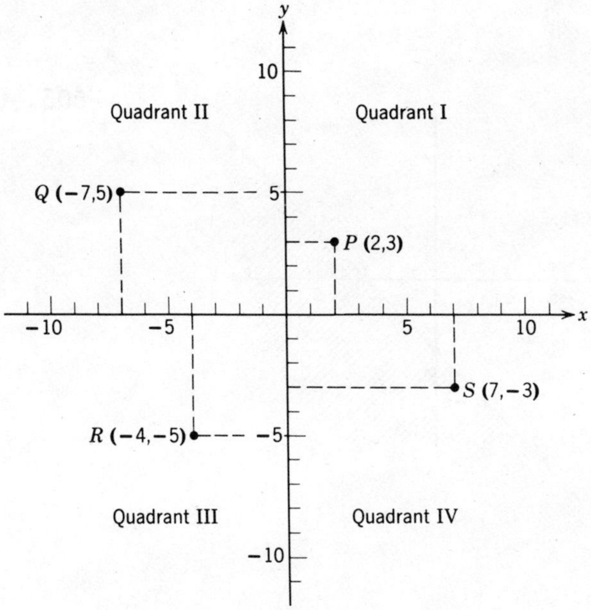

Fig. 4

rectangular

99. [Figure 3] If two line graphs are drawn perpendicular to each other at their origins, they form a rectangular coordinate system. Figure 3 shows a ________________ coordinate system.

axis

100. [Figure 3] The perpendicular line graphs forming a rectangular coordinate system are called axes. The horizontal line is usually called the x-axis and the vertical line is usually called the y-________.

quadrant

101. [Figure 4] The x-axis and y-axis divide the coordinate plane into four regions called quadrants. The quadrants are numbered as illustrated. The region below the x-axis and to the right of the y-axis is the IVth __________.

IV

102. [Figure 4] Point S lies in quadrant ______.

quadrant

103. [Figure 4] Point Q lies in __________ II.

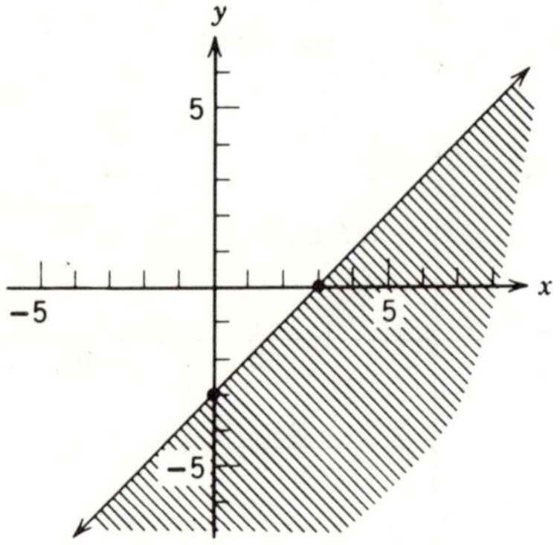

505. Graph $y \leq x - 3$.

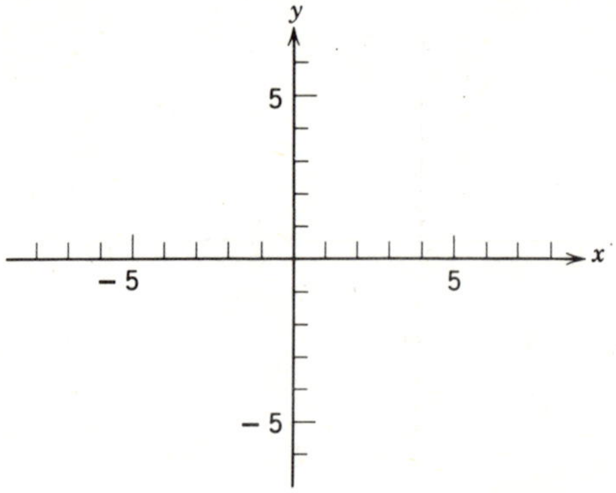

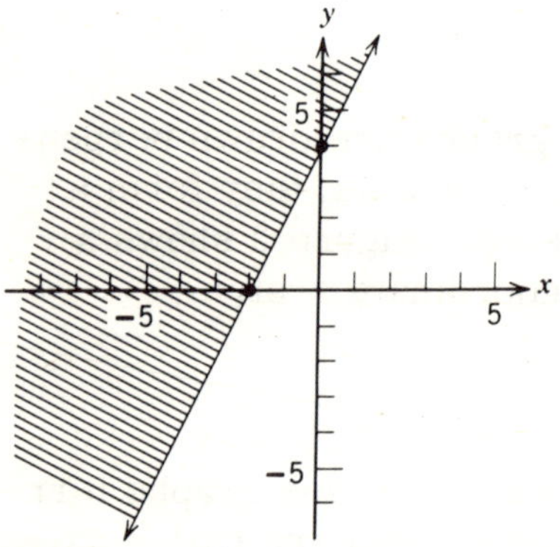

506. Graph $y \geq 2x + 4$.

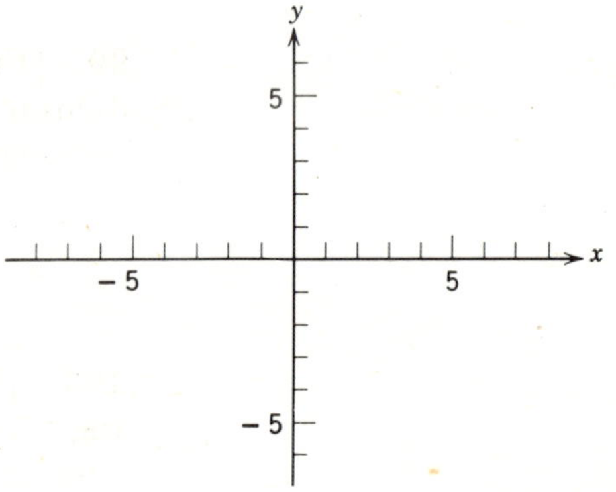

Remark. There are an infinite number of solutions of an equation or an inequality in two variables and therefore it is impossible to list them. Graphing the solution sets provides one means of displaying them. However, there is another way in which such sets can be displayed. The next few frames will introduce you to the symbolism.

solution

507. (0, 2), (1, 3), and (2, 4) are a few of the solutions of $y = x + 2$. The set of all solutions is called the solution set of the equation. Since there are an infinite number of solutions, it would be impossible to list all of the members of the ______________ set.

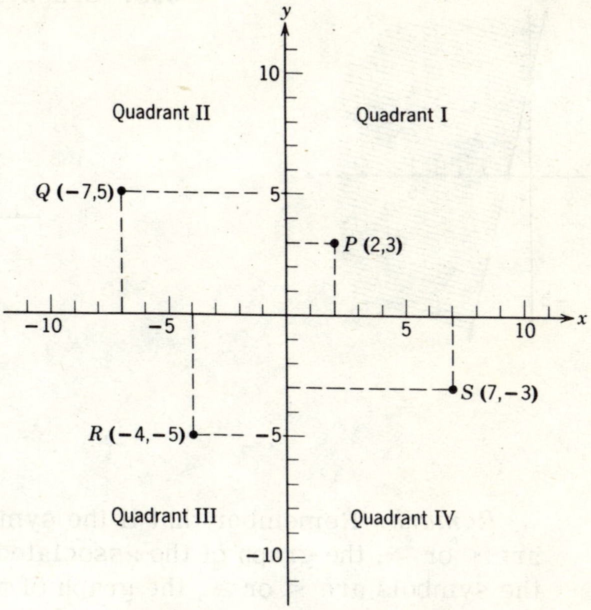

Fig. 4

x

104. [Figure 4] Any point in the plane can be associated with a unique ordered pair of real numbers. Thus, the point designated by P can be associated with the ordered pair (2, 3), where 2 represents the perpendicular distance of the point to the right of the y-axis and 3 represents the perpendicular distance of the point above the_____-axis.

y

105. [Figure 4] Another way to associate the ordered pair (2, 3) with the point P is to think of starting at the origin and moving 2 units to the right along the x-axis, and then, from this point, moving 3 units up parallel to the_____-axis.

above

106. [Figure 4] The point designated by Q can be associated with the ordered pair (−7, 5), where −7 represents the perpendicular distance of the point to the left of the y-axis and 5 represents the perpendicular distance of the point____________ the x-axis.

y-axis

107. [Figure 4] The point Q can be reached by starting at the origin and moving 7 units to the left along the x-axis and, then, from this point, moving 5 units up parallel to the_____________.

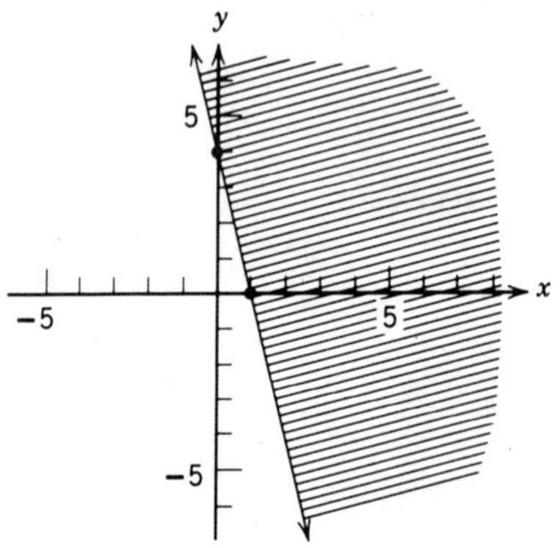

502. Graph $y \geq 4 - 4x$.

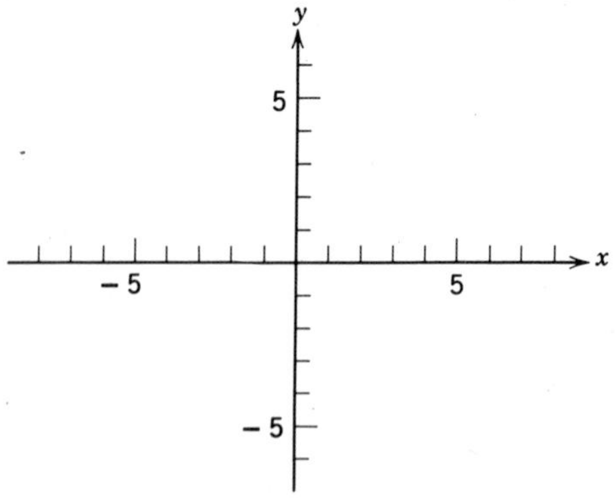

Remark. Remember that if the symbols involved in an inequality are $<$ or $>$, the graph of the associated equation is dashed, while if the symbols are $\leq$ or $\geq$, the graph of the associated equality is shown as a solid line. The next few frames will give you an opportunity to see whether you can graph each type of inequality.

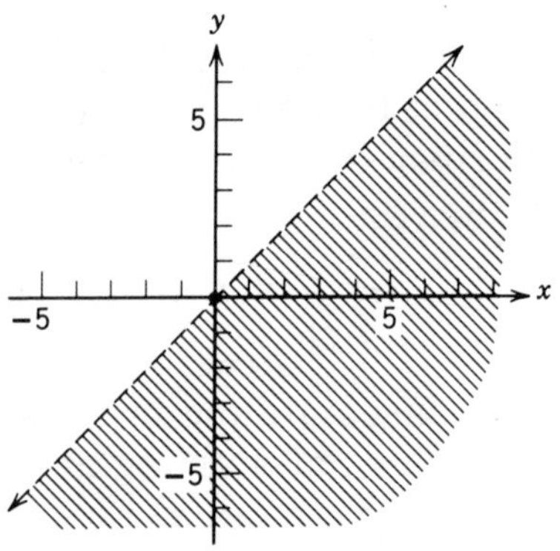

503. Graph $y < x$.

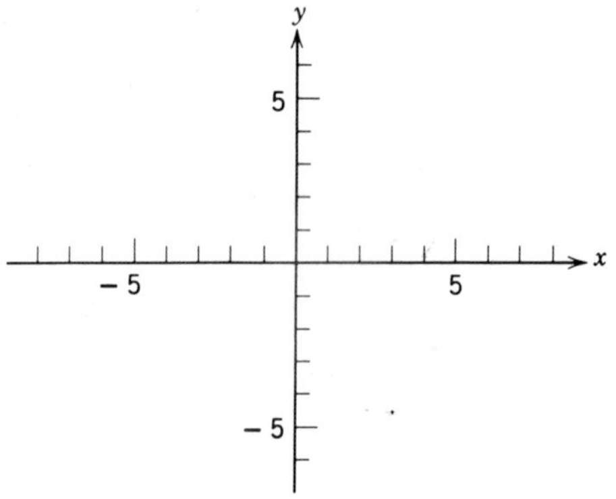

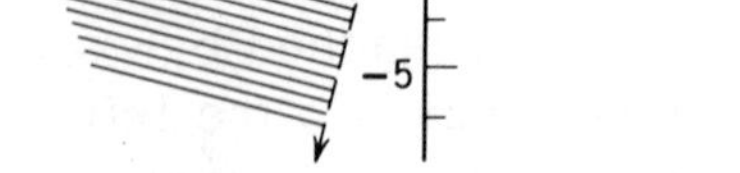

504. Graph $y > 4x + 2$.

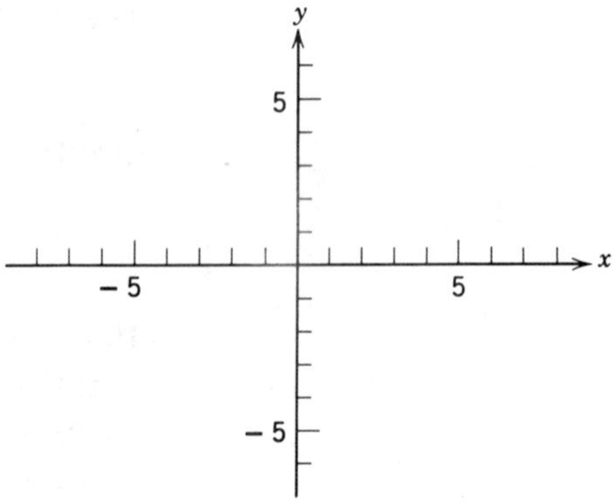

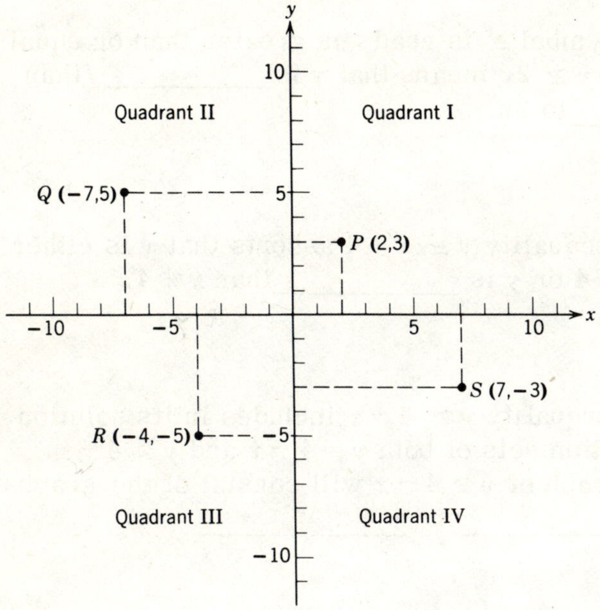

Fig. 4

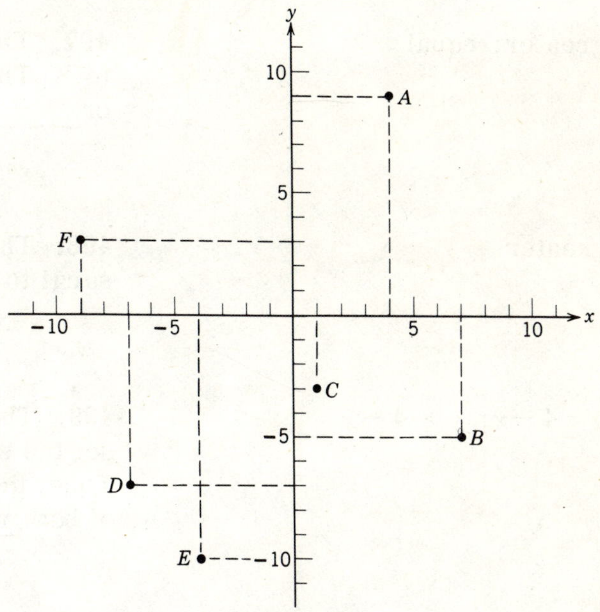

Fig. 5

y-axis; x-axis

108. [Figure 4] The point designated by R can be associated with the ordered pair $(-4, -5)$, where -4 represents the perpendicular distance of the point to the left of the____________ and -5 represents the perpendicular distance of the point below the ____________.

down

109. [Figure 4] The point R can be reached by starting at the origin and moving 4 units to the left along the x-axis and, then, from this point, moving 5 units__________ parallel to the y-axis.

(4, 9)

110. [Figure 5] The point designated by A can be associated with the ordered pair (4,______).

(7, −5)

111. [Figure 5] The point designated by B can be associated with the ordered pair (7,______).

(1, −3)

112. [Figure 5] The point designated by C can be associated with the ordered pair (______, −3).

greater; equal

497. The symbol $\geq$ is read "is greater than or equal to." Thus, $y \geq 2x$ means that y is____________than or__________ to $2x$.

greater

498. The inequality $y \geq x + 4$ asserts that y is either equal to $x + 4$ or y is ____________than $x + 4$.

$y = 4 - x$; $y > 4 - x$

499. The inequality $y \geq 4 - x$ includes in its solution set the solution sets of both $y = 4 - x$ and $y > 4 - x$. Thus, the graph of $y \geq 4 - x$ will consist of the graphs of both $y =$____________ and $y >$____________.

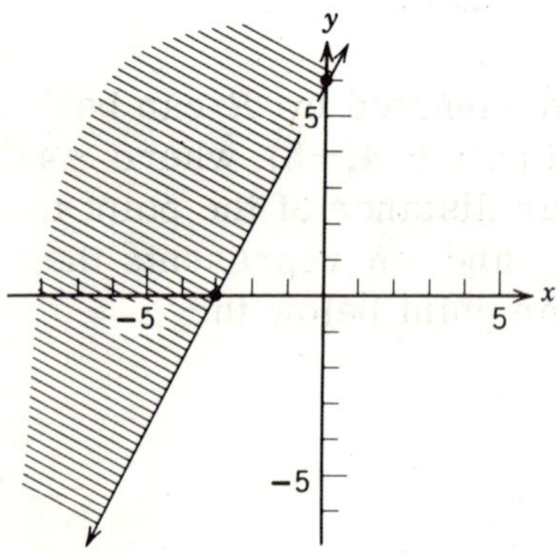

500. Graph the inequality $y \geq 2x + 6$ on the axes provided.

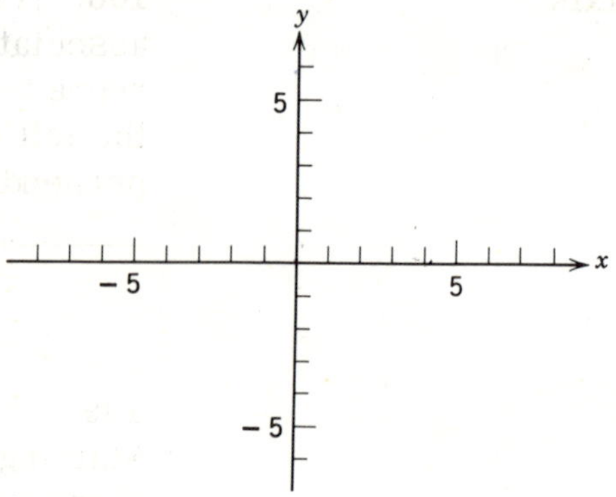

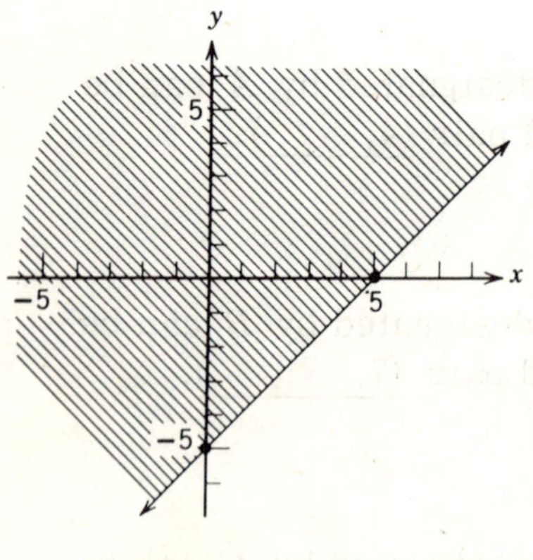

501. Graph the inequality $y \geq x - 5$ on the axes provided.

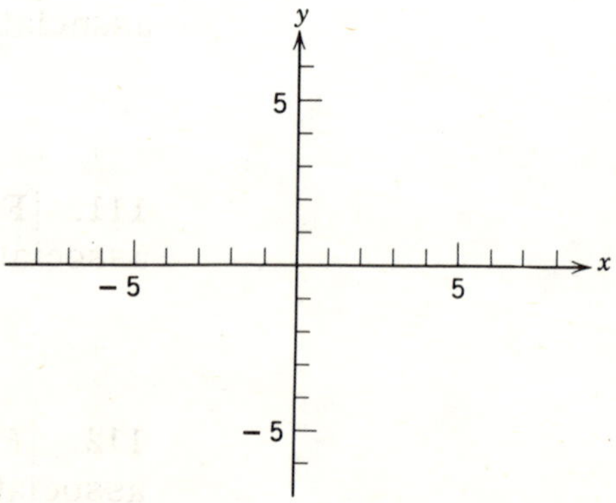

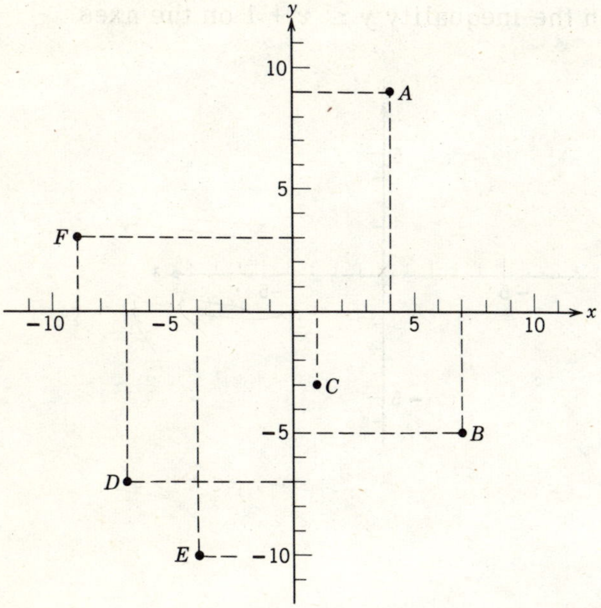

Fig. 5

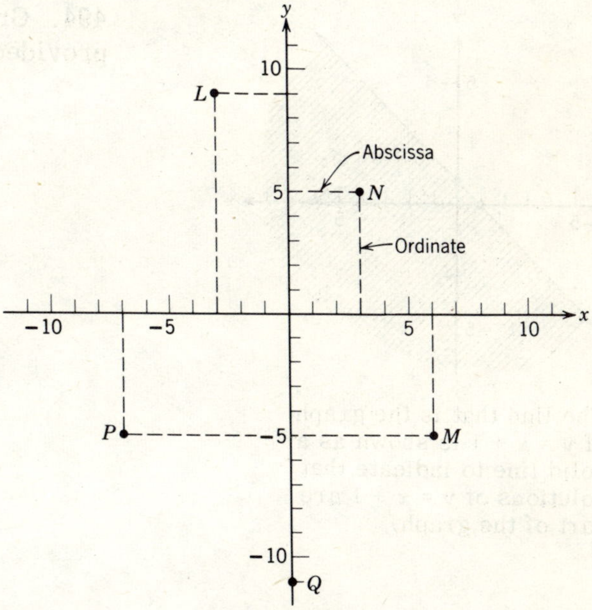

Fig. 6

(−7, −7)

113. [Figure 5] The point designated by D can be associated with the ordered pair (______, −7).

(−4, −10)

114. [Figure 5] The point designated by E can be associated with the ordered pair (______, ______).

(−9, 3)

115. [Figure 5] The point designated by F can be associated with the ordered pair (______, ______).

N

116. [Figure 6] The ordered pair (3, 5) can be associated with the point labeled______.

I

117. [Figure 6] The point labeled N is located in quadrant_____.

L

118. [Figure 6] The ordered pair (−3, 9) can be associated with the point labeled_____.

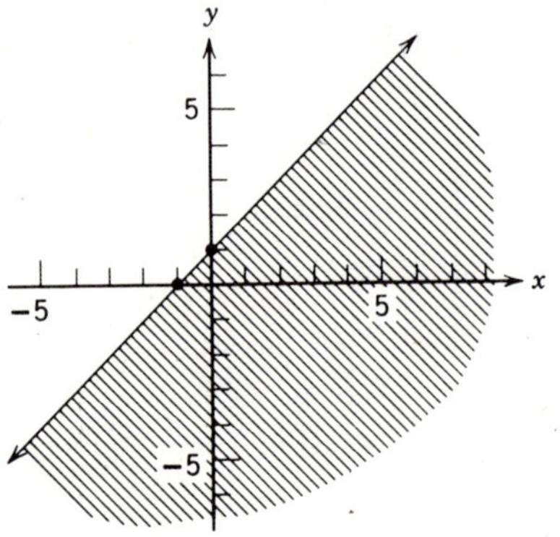

The line that is the graph of $y = x + 1$ is shown as a solid line to indicate that solutions of $y = x + 1$ are part of the graph.

494. Graph the inequality $y \leq x + 1$ on the axes provided.

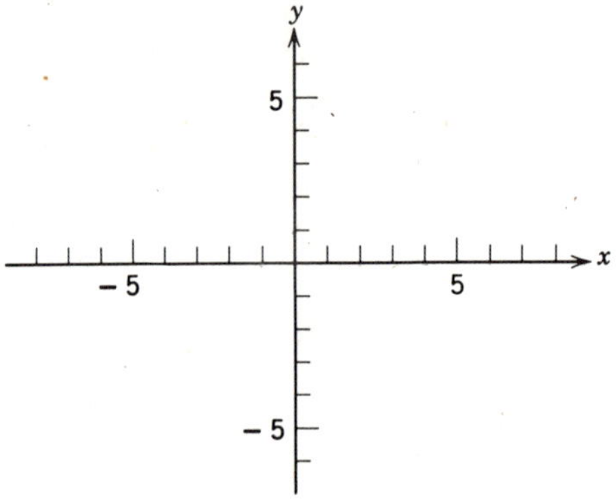

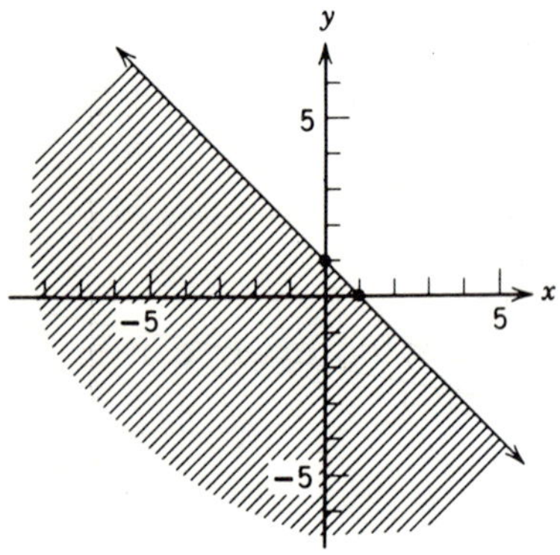

495. Graph the inequality $y \leq 1 - x$ on the axes provided.

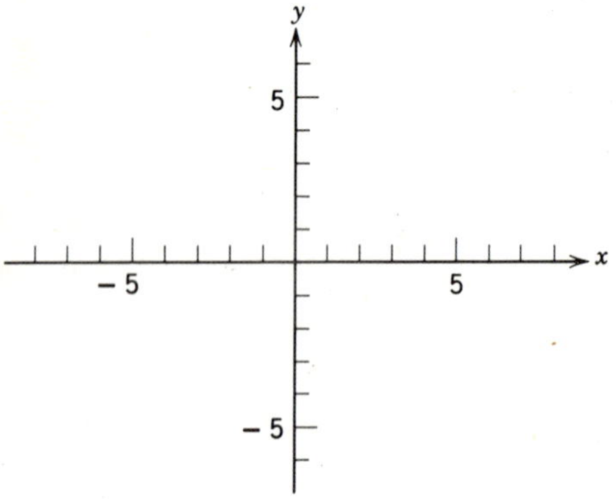

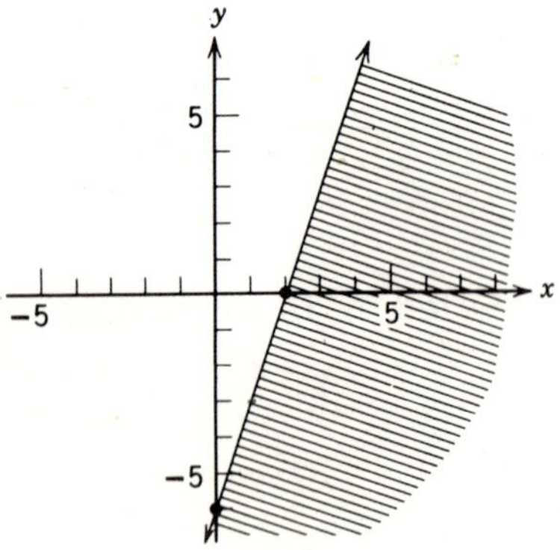

496. Graph $y \leq 3x - 6$.

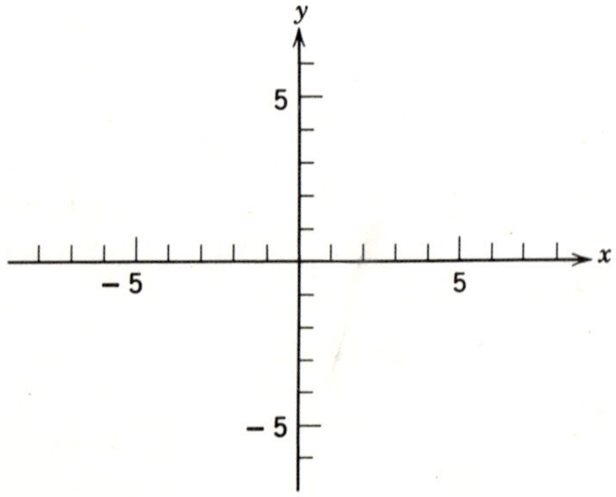

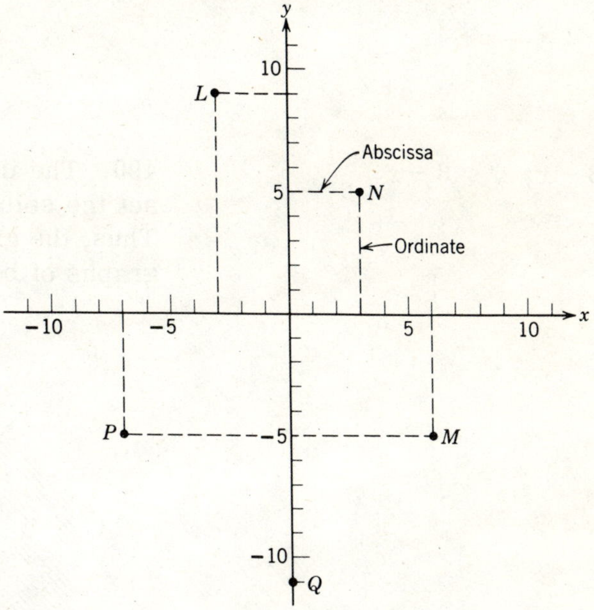

Fig. 6

II

119. [Figure 6] The point labeled L is located in quadrant_____.

M

120. [Figure 6] The ordered pair $(6, -5)$ can be associated with the point labeled_____.

coordinates

121. [Figure 6] The components of an ordered pair associated with a point in the plane are called the coordinates of the point. The components of $(3, 5)$ are the__________________ of the point labeled N.

-3; 9

122. [Figure 6] The coordinates of the point L are ______ and_____.

abscissa

123. The first component of the ordered pair associated with a point in the plane is called the abscissa of the point. For any point P with coordinates designated by x and y, the number represented by x is the_________________.

$y = 3 - x$; $y < 3 - x$

490. The inequality $y \leq 3 - x$ includes in its solution set the solution sets of both $y = 3 - x$ and $y < 3 - x$. Thus, the graph of $y \leq 3 - x$ will consist of the graphs of both $y =$ ________ and $y <$ ________.

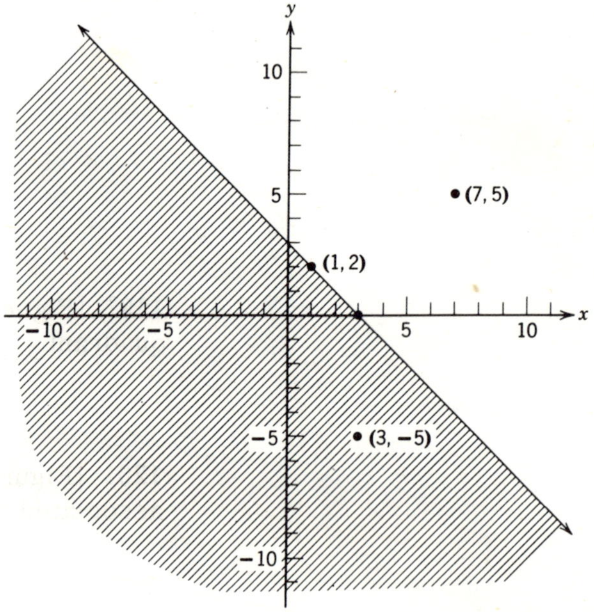

Fig. 32

is

491. [Figure 32] This figure shows the graph of $y \leq 3 - x$. The line $y = 3 - x$ is shown as a solid line indicating that the line is a part of the graph of the solution set. The ordered pair (1, 2) (is/is not) in the solution set.

is

492. [Figure 32] The ordered pair (3, −5) (is/is not) in the solution set of $y \leq 3 - x$.

is not

493. [Figure 32] The ordered pair (7, 5) (is/is not) in the solution set of $y \leq 3 - x$.

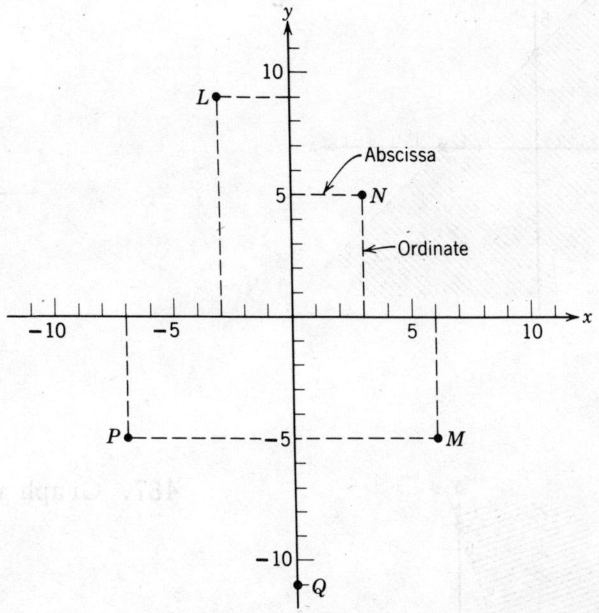

Fig. 6

ordinate

124. The second component of an ordered pair associated with a point in the plane is called the ordinate of the point. For any point P with coordinates designated by x and y, the number represented by y is the____________________.

ordinate

125. [Figure 6] Referring to point M, 6 is the abscissa and -5 is the____________________.

abscissa

126. [Figure 6] Referring to point N, 3 is the ____________________and 5 is the ordinate.

-3

127. [Figure 6] The ordinate of point L is 9 and the abscissa is_______.

-7; -5

128. [Figure 6] The abscissa of point P is________ and the ordinate is________.

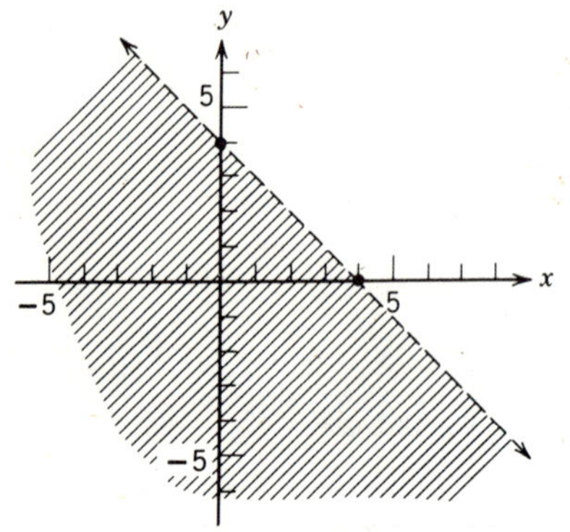

486. Graph $y < 4 - x$.

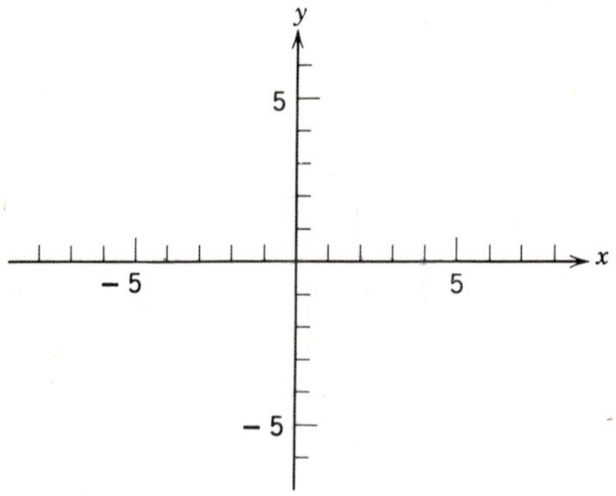

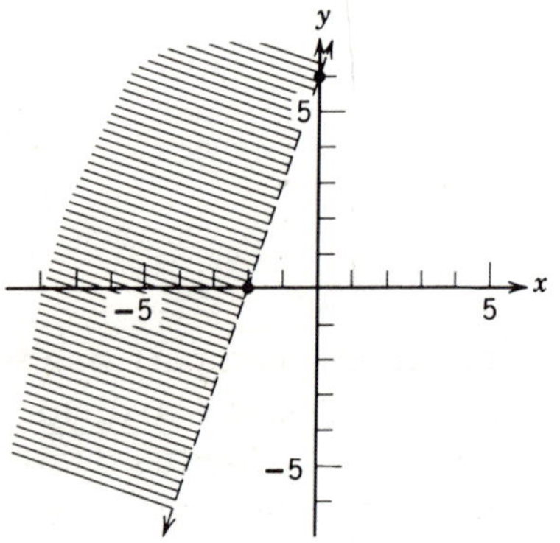

487. Graph $y > 6 + 3x$.

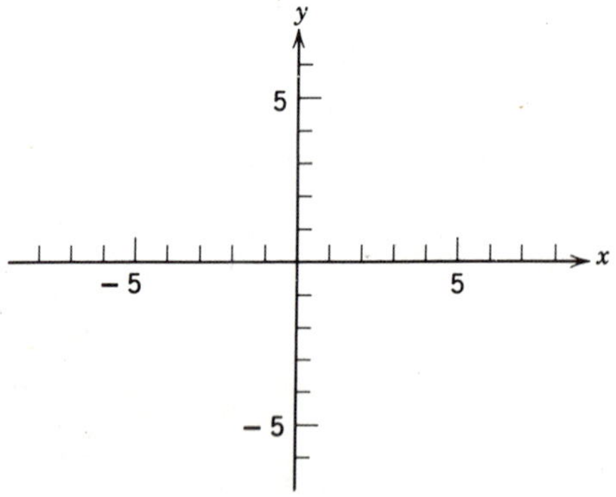

Remark. If the inequality we are graphing excludes the solutions of the associated equation, the graph of the equation is shown as a dashed line. However, sometimes we have inequalities in which the relationship includes the case where the members are equal as well as the case where they are unequal.

less; equal

488. The symbol "$\leq$" means "is less than or equal to." Thus, $y \leq x + 1$ is read "y is ________ than or ________ to $x + 1$."

less

489. The inequality $y \leq x + 3$ asserts that y is either equal to $x + 3$ or y is ________ than $x + 3$.

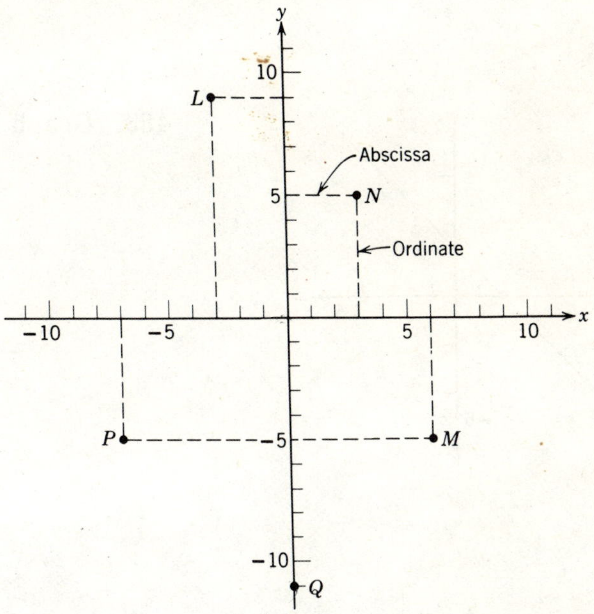

Fig. 6

y

129. [Figure 6] The abscissa of a point is the perpendicular distance from the_____-axis to the point.

x

130. [Figure 6] The ordinate of a point is the perpendicular distance from the____-axis to the point.

abscissa; ordinate

131. [Figure 6] Referring to point Q, 0 is the __________________and -11 is the ________________.

coordinate

132. An ordered pair can be associated with a point in the plane by means of a rectangular_____________ system.

Cartesian

133. A rectangular coordinate system is also called a Cartesian coordinate system after its inventor, the great French mathematician and philosopher Rene Descartes. Another name for a rectangular coordinate system is__________________coordinate system.

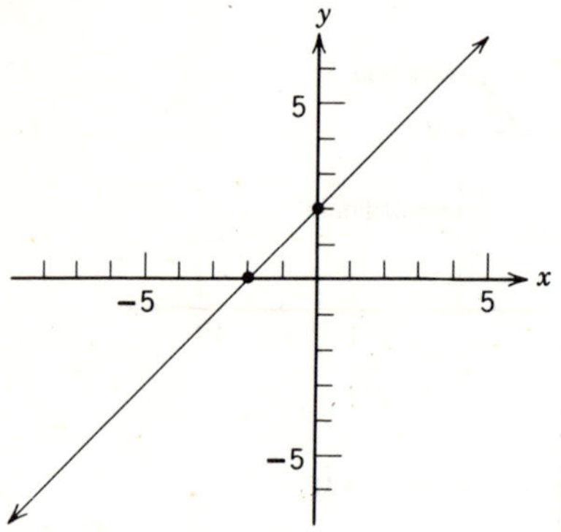

483. Graph $y = x + 2$.

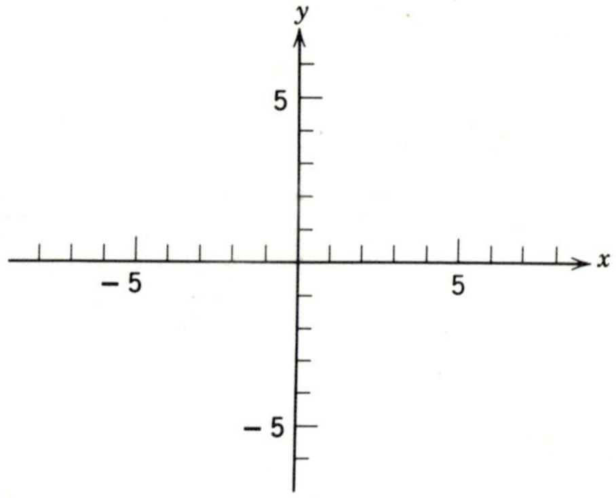

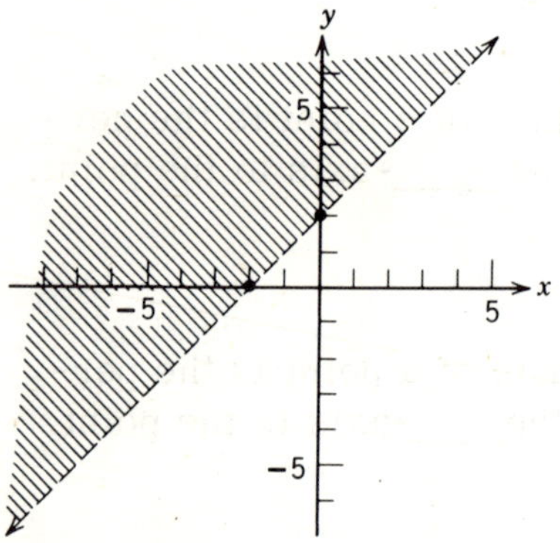

484. Graph $y > x + 2$.

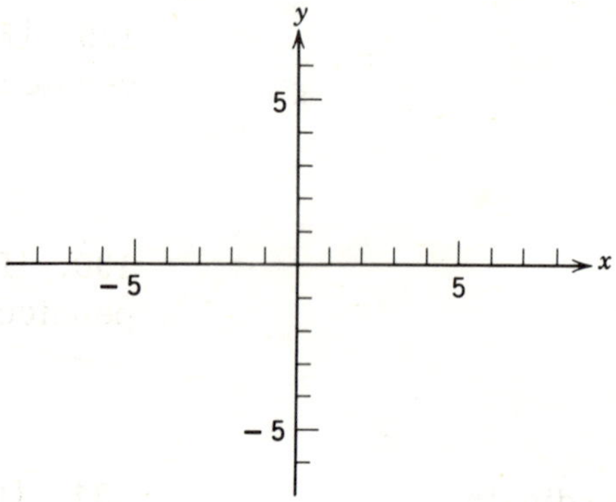

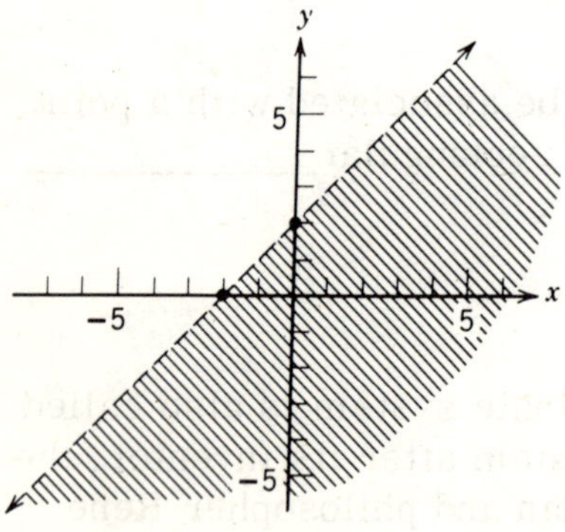

485. Graph $y < x + 2$.

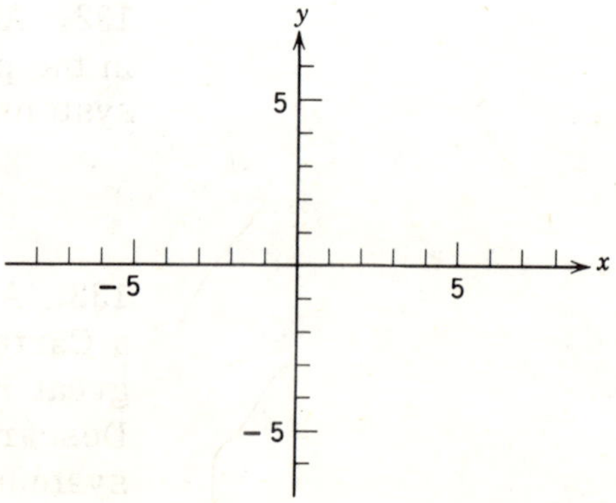

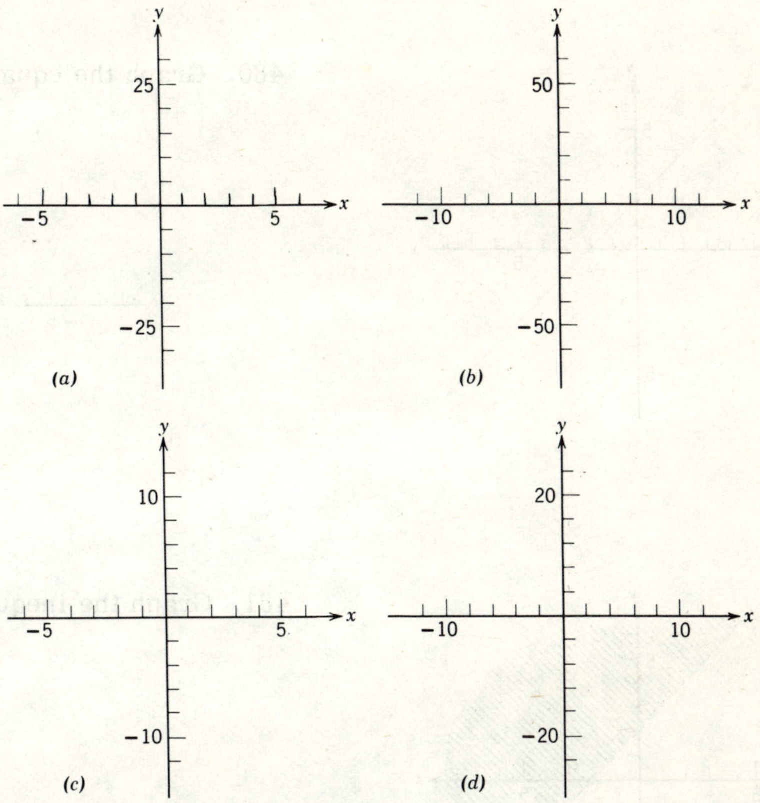

Fig. 7

1

134. [Figure 7a] In this coordinate system each division on the x-axis represents____ unit(s) and each division on the y-axis represents 5 units.

2; 10

135. [Figure 7b] In this coordinate system each division on the x-axis represents____ unit(s) and each division on the y-axis represents____ unit(s).

1; 2

136. [Figure 7c] In this coordinate system each division on the x-axis represents____ unit(s) and each division on the y-axis represents____ unit(s).

2; 4

137. [Figure 7d] In this coordinate system each division on the x-axis represents____ unit(s) and each division on the y-axis represents____ unit(s).

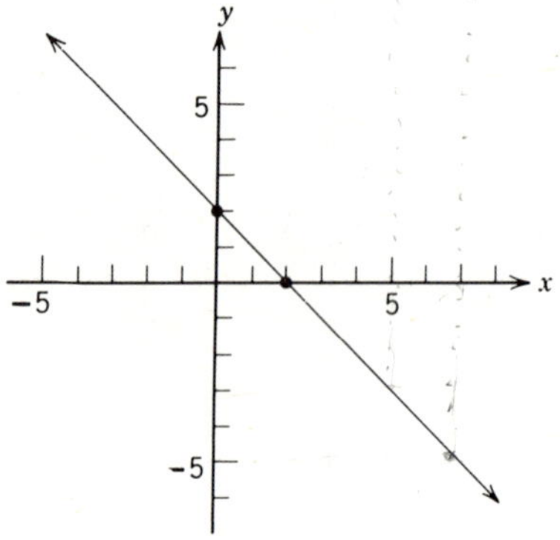

480. Graph the equality $y = 2 - x$.

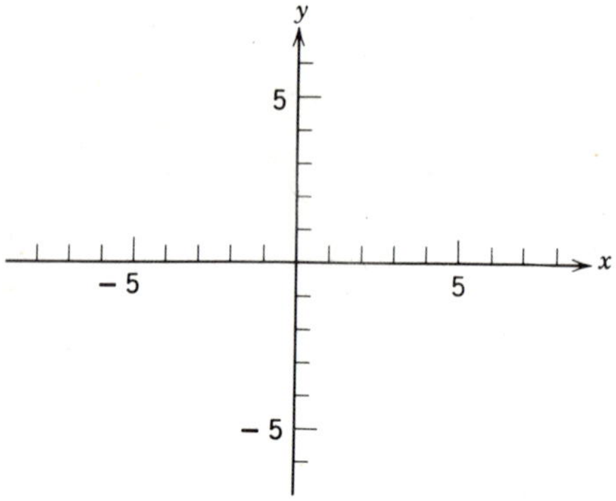

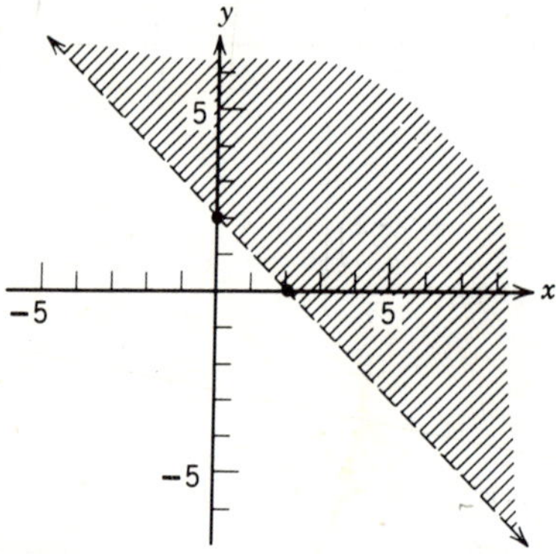

The line is dashed to indicate that solutions of $y = 2 - x$ are not part of the graph.

481. Graph the inequality $y > 2 - x$.

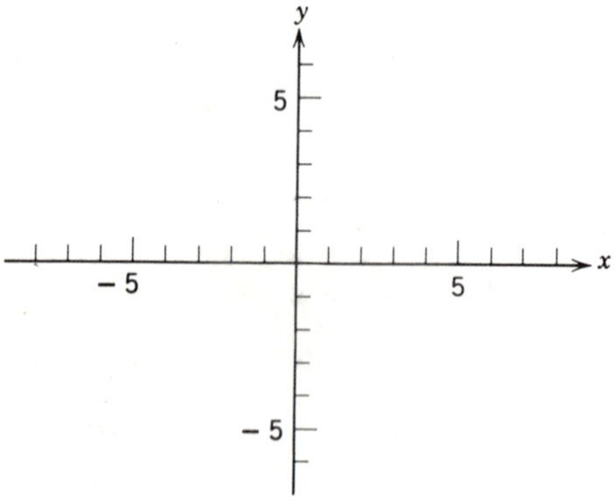

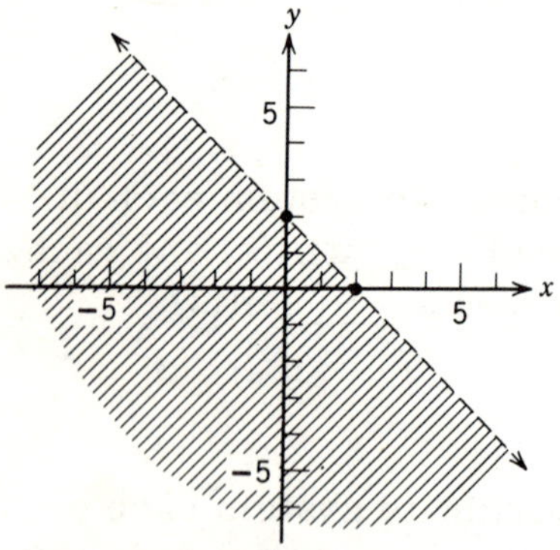

482. Graph the inequality $y < 2 - x$.

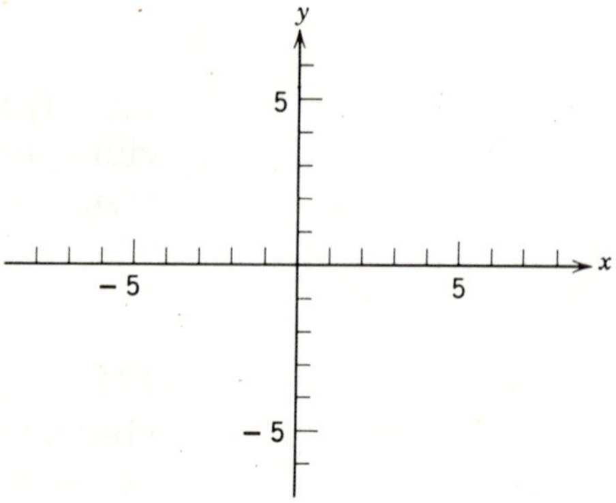

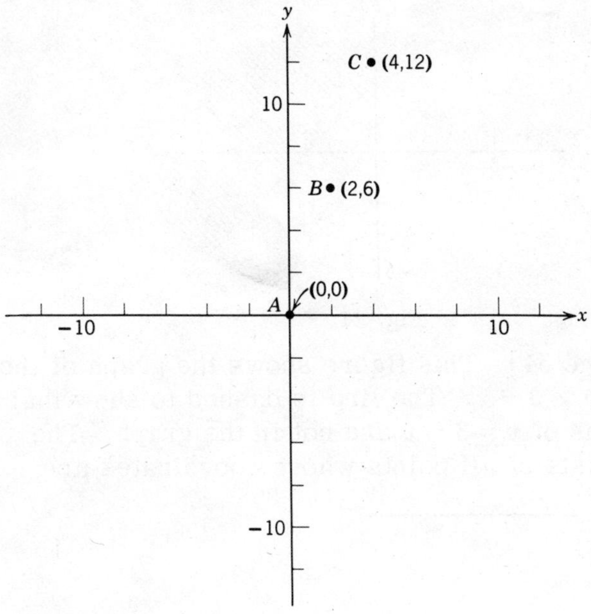

Fig. 8

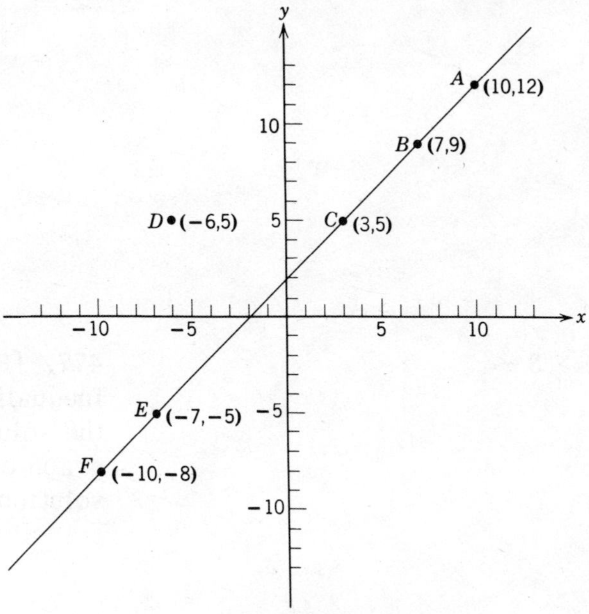

Fig. 9

(0, 2); (5, 7); (10, 12)

138. Recall that the solution of a first-degree equation in two variables is an ordered pair of numbers. Some solutions of $y = x + 2$ are (0, ____), (5, ____), and (10, ____).

(0, 1); (2, 7); (5, 16)

139. Some solutions of $y = 3x + 1$ are (0, ____), (2, ____), and (5, ____).

graph

140. [Figure 8] Some solutions of $y = 3x$ are (0, 0), (2, 6), and (4, 12). These ordered pairs can be associated with the points labeled A, B, and C respectively. The points are called the graphs of the solutions. The point corresponding to (−2, −6) is also the ________ of a solution of the equation $y = 3x$.

is

(10, 12) is a solution of $y = x + 2$.

141. [Figure 9] The point labeled A (is/is not) the graph of a solution of the equation $y = x + 2$.

is

142. [Figure 9] The point labeled B (is/is not) the graph of a solution of the equation $y = x + 2$.

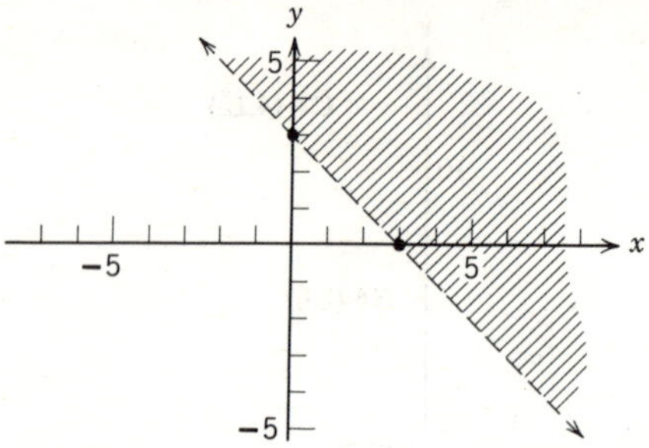

Fig. 31

$y > 3 - x$

477. [Figure 31] This figure shows the graph of the inequality $y > 3 - x$. The line is dashed to show that the solutions of $y = 3 - x$ are not in the graph. The graph consists of all points whose coordinates are solutions of______________.

For any x, all values of y that make $y < 4 + x$ true, are associated with points that lie below the line $y = 4 + x$.

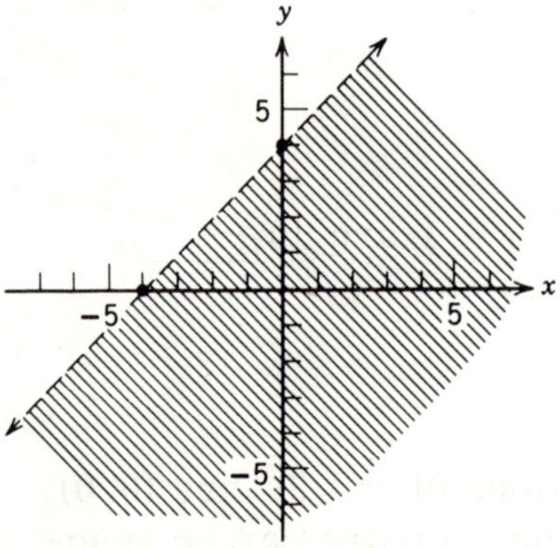

478. Given the graph of the equation $y = 4 + x$, shade the area that consists of all points whose coordinates are solutions of the inequality $y < 4 + x$.

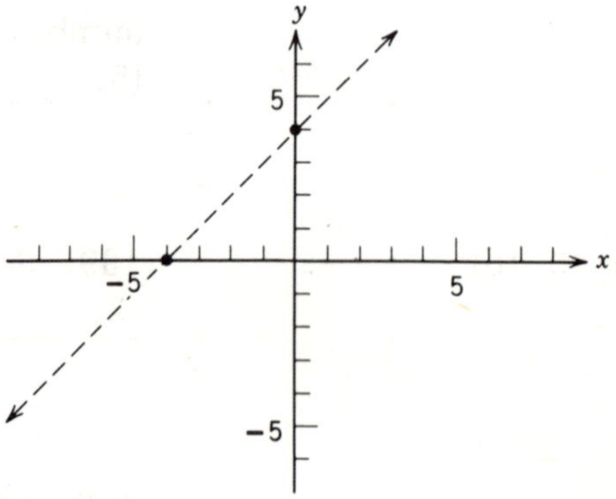

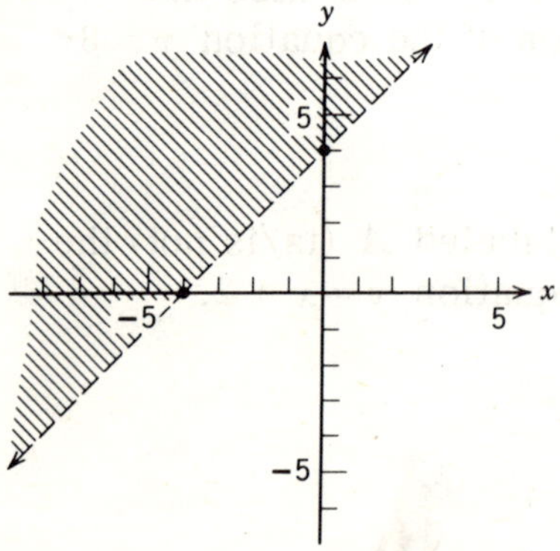

479. Given the graph of the equation $y = 4 + x$, shade the area that consists of all points whose coordinates are solutions of the inequality $y > 4 + x$.

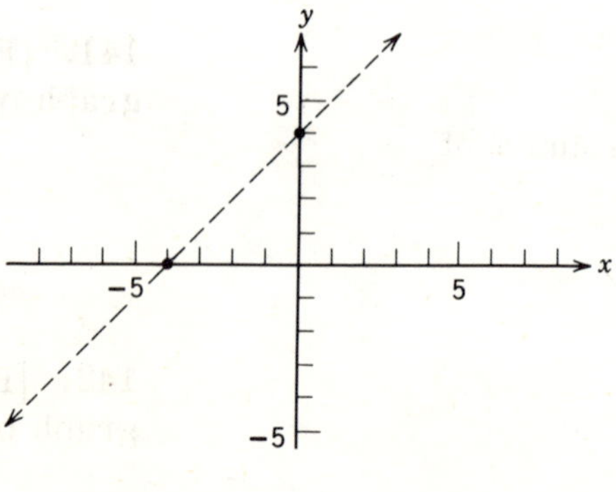

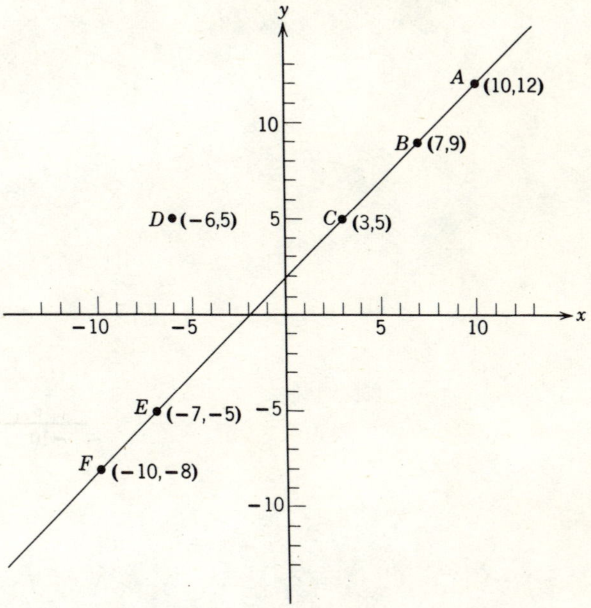

Fig. 9

is

143. [Figure 9] The point labeled C (is/is not) the graph of a solution of the equation $y = x + 2$.

is not

144. [Figure 9] The point labeled D (is/is not) the graph of a solution of the equation $y = x + 2$.

$(-6, 5)$ is not a solution of $y = x + 2$.

is

145. [Figure 9] The point labeled E (is/is not) the graph of a solution of the equation $y = x + 2$.

is

146. [Figure 9] The point labeled F (is/is not) the graph of a solution of the equation $y = x + 2$.

straight

147. [Figure 9] The graphs of some solutions of $y = x + 2$ as shown here appear to lie on a straight line. It can be proved that the graphs of all solutions of a given first-degree equation in two variables lie on a given straight line. The graph of the solution set of $y = x + 2$, therefore, is a ________________ line.

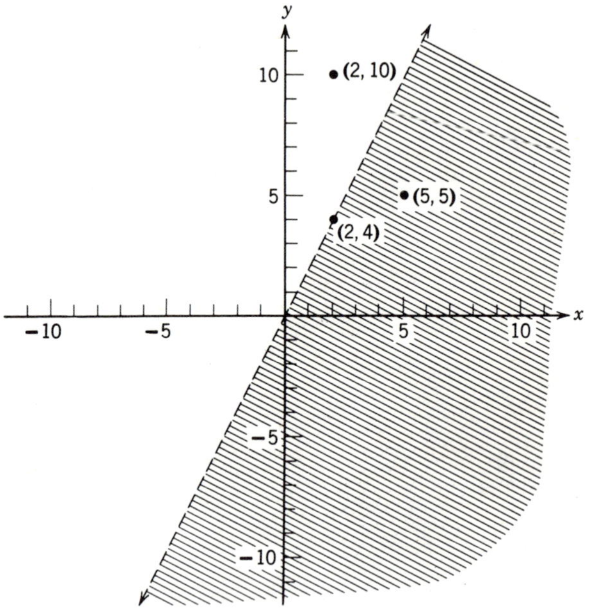

Fig. 30

is

473. [Figure 30] The shaded portion of this figure shows the graph of the solution set of $y < 2x$. (5, 5) lies in the shaded portion of the graph. (5, 5) (is/is not) a solution of $y < 2x$.

below

474. [Figure 30] The entire region below the line $y = 2x$ is shaded because for each value of x, $y < 2x$ is satisfied by the coordinates of every point (below/above) the line that is the graph of $y = 2x$.

is not

475. [Figure 30] (2, 10) does not lie in the shaded portion of the graph and (is/is not) a solution of $y < 2x$.

is not

Substituting 2 for x and 4 for y yields $4 < 4$, which is not true.

476. [Figure 30] The line that is the graph of $y = 2x$ is shown by dashes to indicate that the line is not a part of the graph of $y < 2x$. The point whose coordinates are (2, 4) lies on the line. (2, 4)(is/is not) a solution of $y < 2x$.

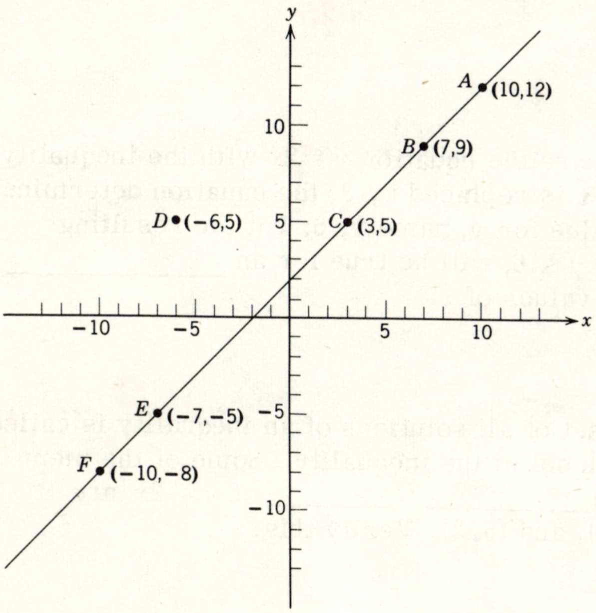

Fig. 9

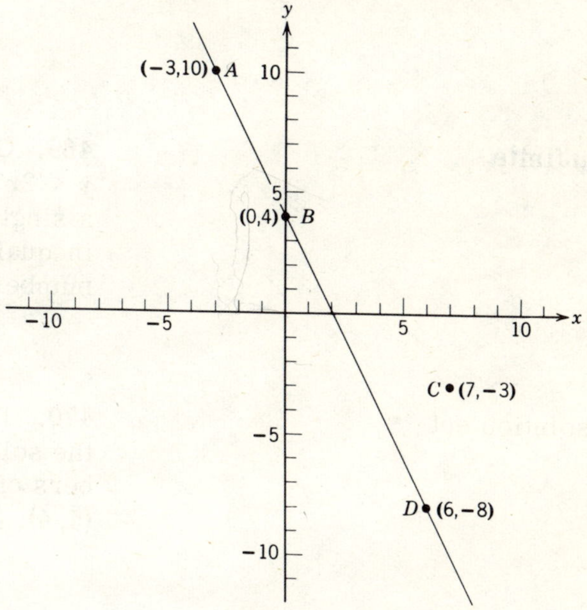

Fig. 10

does not

148. The graphs of any ordered pair that is not a solution of a given equation does not lie on the straight line that is the graph of the equation. In Figure 9, the ordered pair (−6, 5) is not a solution of $y = x + 2$ and it is clear that its graph (does/does not) lie on the line.

is

The ordered pair (−3, 10) is a solution. $10 = 4 - 2(-3)$ or $10 = 10$ is true.

149. [Figure 10] The point labeled A (is/is not) the graph of a solution of $y = 4 - 2x$.

is

150. [Figure 10] The point labeled B (is/is not) the graph of a solution of $y = 4 - 2x$.

is not

151. [Figure 10] The point labeled C (is/is not) the graph of a solution of $y = 4 - 2x$.

is

152. [Figure 10] The point labeled D (is/is not) the graph of a solution of $y = 4 - 2x$.

infinite

469. Compare the equation $y = 2x$ with the inequality $y < 2x$. If x is replaced by 3, the equation determines a single value for y, namely, 6; but the resulting inequality, $y < 6$, will be true for an ____________ number of values of y.

solution set

470. The set of all solutions of an inequality is called the solution set of the inequality. Some of the members of the__________ ________ of $y < 2x$ are (5, 4), (5, 3), and (5, 2). Verify this.

solution set

471. All the ordered pairs of numbers that are solutions to an inequality in two variables make up the ____________ ________of the inequality.

$y = 2x$

472. [Figure 29] The graph of $y = 2x$ consists of all points on the line. The coordinates of points on the line, for example (2, 4), (3, 6), and (4, 8), are solutions of the equation____ = ________.

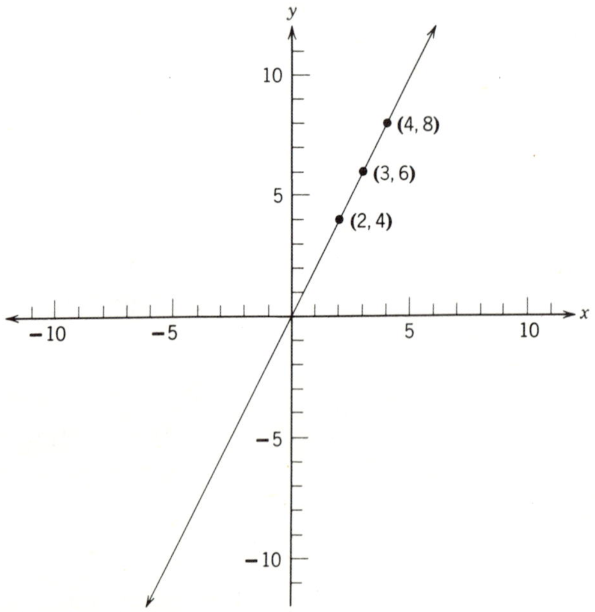

Fig. 29

straight

153. The solution set of a first-degree equation in two variables, such as $y = x + 5$, contains an infinite number of ordered pairs of numbers. The graph of the solution set would be an infinite set of points, all of which lie on a ______________ line.

infinite; line

154. The graph of the solution set of $4x - y = 7$ is an ______________ set of points, all of which lie on a straight ___________.

graph
Or, line.

155. [Figure 11] The straight line on which the graphs of the solutions of $y = x + 5$ lie is called the graph of the equation. Obviously, only part of the line can be displayed, and an arrowhead is placed at either end to indicate that the ____________ continues in both directions indefinitely.

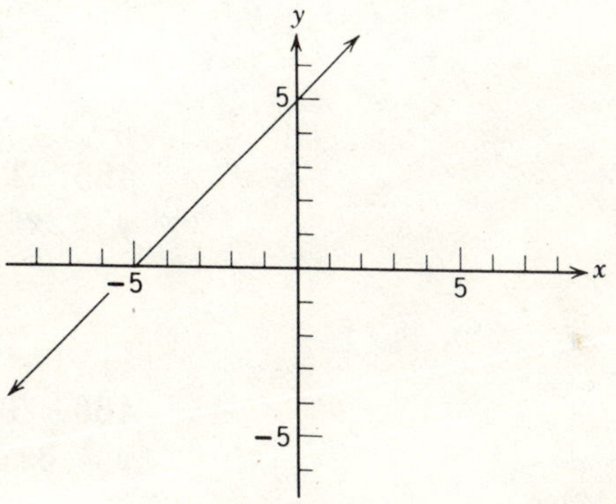

Fig. 11

graph

156. The straight line on which the solutions of $x - 2y = 6$ lie is called the ____________ of the equation.

linear

157. Since the graphs of all first-degree equations in two variables are straight lines, such equations are also called linear equations. $y = x + 5$ is a __________ equation.

linear

158. $2x + 3y = 5$ is a first-degree or ______________ equation.

3

461. Inequalities in two variables such as $y > x + 1$ have an infinite number of solutions. If x represents 2, the inequality can be written $y > 2 + 1$ or $y > 3$, which is true as long as y is greater than_____.

greater

462. The ordered pair $(2, y)$ will be a solution of the inequality $y > x + 1$ as long as y is ___________ than 3.

6

463. When $x = 5$, the inequality $y > x + 1$ can be written $y > 6$, which is true as long as y is greater than_____.

greater

464. The ordered pair $(5, y)$ will be a solution of $y > x + 1$ as long as y is____________ than 6.

less

465. The ordered pair $(1, y)$ will be a solution of $y < 2x + 1$ as long as y is________ than 3.

-10

Substituting -4 for x yields $y < -10$.

466. The ordered pair $(-4, y)$ will be a solution of $y < 3x + 2$ as long as y is less than________.

less than

Substituting 3 for x yields $y < 0$.

467. The ordered pair $(3, y)$ will be a solution of $y < 3 - x$ as long as y is__________ _______ 0.

less than

468. Compare the equation $y = x + 1$ with the inequality $y < x + 1$. If x is replaced by 3 in the equation, the result is $y = 4$ and the ordered pair $(3, 4)$ is a solution. However, if x is replaced by 3 in $y < x + 1$, the result is $y < 4$, which is true for all values of y__________ ______4.

first-degree

159. The expressions "linear equation" and "__________ __________ equation" mean the same thing.

straight

160. The graph of a linear equation is a __________ line.

points

161. The graph of a linear equation is a straight line and is determined by any two points on the line. The graph of $y = 2x - 3$ can be determined by any two __________ whose coordinates satisfy the equation.

Two

162. The graph of $x + 6y = 12$ is a straight line. __________ points are necessary to determine the line.

two

163. The graph of $x - 3y = 6$ can be determined by any __________ points whose coordinates satisfy the equation.

(0, 5); (5, 0)

164. [Figure 12] Two solutions of $y = 5 - x$ are (0, ____) and (5, ____). The graphs of the two solutions are shown.

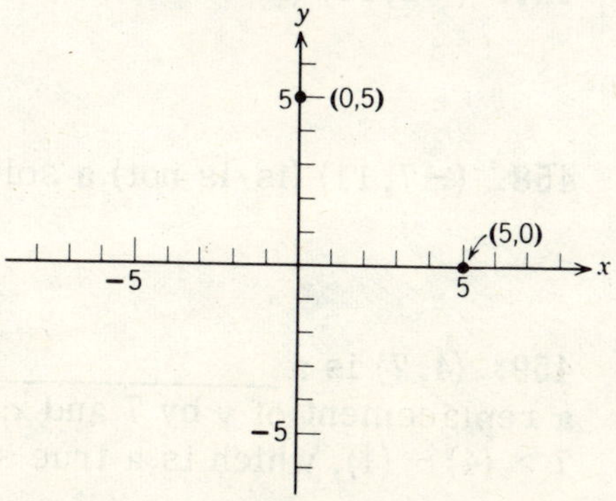

Fig. 12

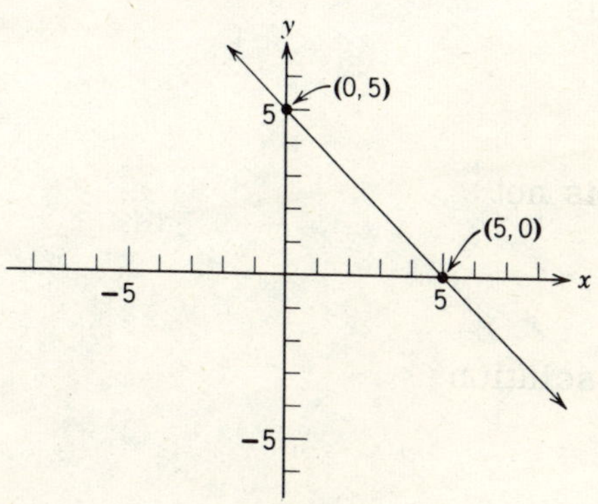

Fig. 13

graph

165. [Figure 13] The line drawn through (0, 5) and (5, 0) is the __________ of the equation $y = 5 - x$.

inequality

452. Expressions involving the symbols $<$ and $>$ are called inequalities. Thus, $y < 2x$ is an example of an________________.

true

453. Inequalities involving two variables have ordered pairs of numbers as solutions. Thus, (2, 3) is a solution of $x + y < 7$, because when x is replaced by 2 and y by 3, the resulting statement, $2 + 3 < 7$, is (true/false).

true

454. (4, 3) is a solution of the inequality $x + y > 6$ because a replacement of x by 4 and y by 3 yields $(4) + (3) > 6$, which is a__________statement.

is not

455. (4, 1) is not a solution of the inequality $x + y > 6$ because a replacement of x by 4 and y by 1 yields $(4) + (1) > 6$, which (is/is not) a true statement.

is

456. (0, 8) (is/is not) a solution of $x + y > 6$.

is

457. (−2, 12) (is/is not) a solution of $x + y > 6$.

is not

458. (−7, 11) (is/is not) a solution of $x + y > 6$.

solution

459. (4, 7) is a______________of $y > x + 1$ because a replacement of y by 7 and x by 4 yields $7 > (4) - (1)$, which is a true statement.
(+)

is not

460. (3, 0) (is/is not) a solution of $y > x + 1$.

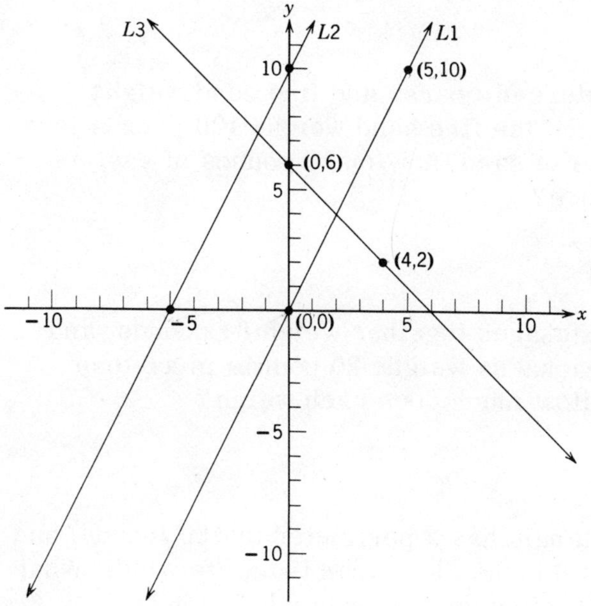

Fig. 14

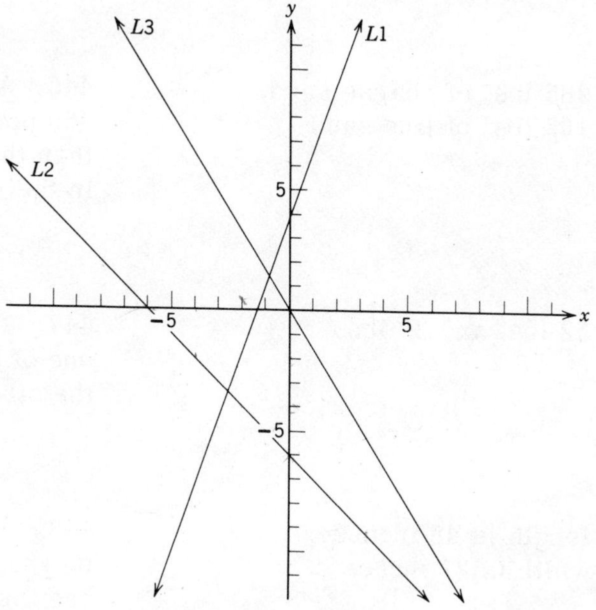

Fig. 15

(0, 6); (4, 2)

166. Two solutions of $x + y = 6$ are (0, ______) and (4, ______).

L-3

167. [Figure 14] Since two solutions of $x + y = 6$ are (0, 6) and (4, 2), the graph of the equation $x + y = 6$ appears as line *L*-______.

(0, 0); (5, 10); *L*-1

168. [Figure 14] Two solutions of $y - 2x = 0$ are (0, ______) and (5, ______). The graph of the equation $y - 2x = 0$ appears as line *L*-______.

(0, 10); (−5, 0); *L*-2

169. [Figure 14] Two solutions of $y - 2x = 10$ are (0, ______) and (−5, ______). The graph of $y - 2x = 10$ appears as line *L*-______.

L-2

Did you find two solutions of the equation, graph the solutions, and find the line determined by these points?

170. [Figure 15] The graph of the equation $x + y = -6$ appears as line *L*-______.

285 lbs. of coarse sand;
165 lbs. of fine sand

446. A mixture of coarse and fine sand weighs 450 pounds. If the fine sand weighs 120 pounds less than the coarse sand, how many pounds of each is in the mixture?

52 lbs. and 32 lbs.

447. Two packages together weigh 84 pounds, and one of the packages weighs 20 pounds more than the other. How much does each weigh?

length is 48 inches;
width is 27 inches

448. A rectangle has a perimeter of 150 inches, and its length is 6 inches less than twice its width. What are the dimensions of the rectangle?

33 ft. and 27 ft.

449. A 60-foot rope is cut into two pieces, one of which is 6 feet shorter than the other. How long is each piece?

Remark. There are many other types of situations in which two equations in two variables can be used to express relationships between things. If you continue to study algebra, or if you take courses in one or another of the physical sciences, you will encounter places where this technique is useful.

As a final topic for this unit, we are going to examine inequalities in two variables. You will recall that an inequality is a mathematical sentence involving signs such as $>$, $<$, $\geq$, $\leq$.

less

450. Recall that the symbol "$<$" means "is less than." Thus, if x and y represent numbers, $y < x$ asserts that y represents a number __________ than that represented by x.

greater

451. The symbol "$>$" means "is greater than." Thus, $y > x + 2$ asserts that y represents a number that is ____________ than the number represented by the sum of x and 2.

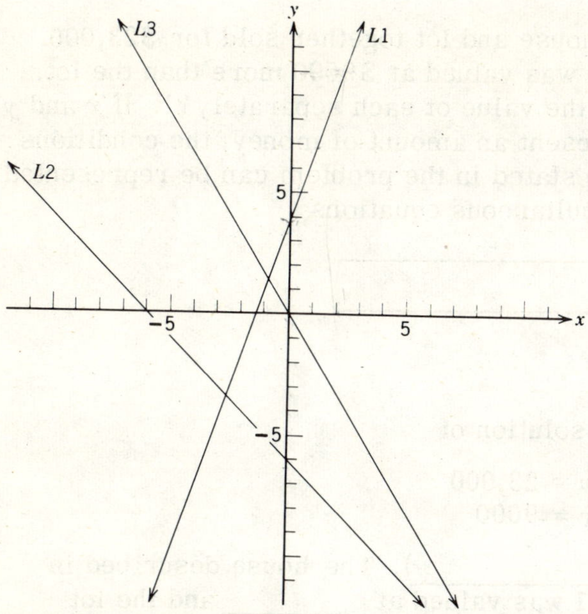

Fig. 15

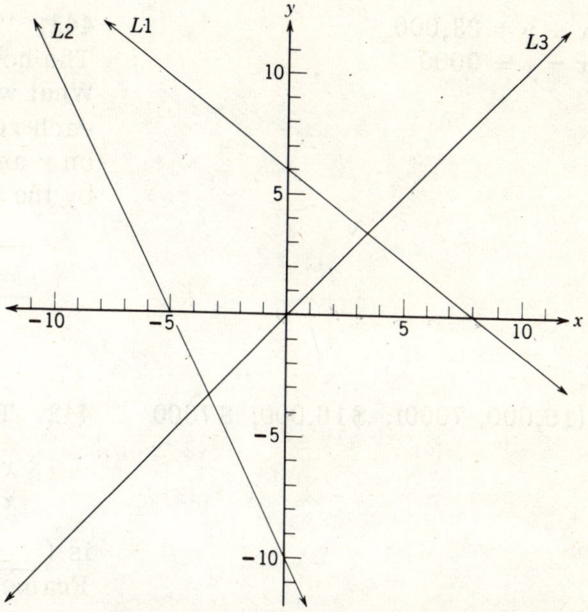

Fig. 16

L-1

171. [Figure 15] The graph of the equation $3x - y = -4$ appears as line L-_____.

L-1

172. [Figure 16] The graph of $3x + 4y = 24$ appears as line L-_____.

L-3

173. [Figure 16] The graph of $y = x$ appears as line L-_____.

(0, −3); (3, 0)

174. Using 0 and 3 as first components, two solutions of the equation $y = x - 3$ are (____, ____) and (____, ____).

$x + y = 23{,}000$
$x - y = 9000$

441. "A house and lot together sold for \$23,000. The house was valued at \$9000 more than the lot. What was the value of each separately?" If x and y each represent an amount of money, the conditions on x and y stated in the problem can be represented by the simultaneous equations

_________ = _________

_________ = _________.

(16,000, 7000); \$16,000; \$7000

442. The solution of

$$x + y = 23{,}000$$
$$x - y = 9000$$

is (_______, _______). The house described in Frame 441 was valued at _________ and the lot was valued at _________.

$x + y = 300$
$x - y = 12$

443. "Two trains left towns A and B, 300 miles apart, at the same time and proceeded toward each other on parallel tracks. At the time they met, the train from A had traveled 12 miles farther than the train from B. How many miles from A were the trains when they met?" Using variables x and y to represent distances from towns A and B respectively, represent the conditions on x and y stated in the problem.

(156, 144); 156 miles

The train from A traveled farther than the train from B.

444. The solution of

$$x + y = 300$$
$$x - y = 12$$

is (_______, _______). Therefore, the trains referred to in Frame 443 met at a point _______ _________ from town A.

-24 and -16

445. The sum of two numbers is -40 and their difference is -8. What are the numbers?

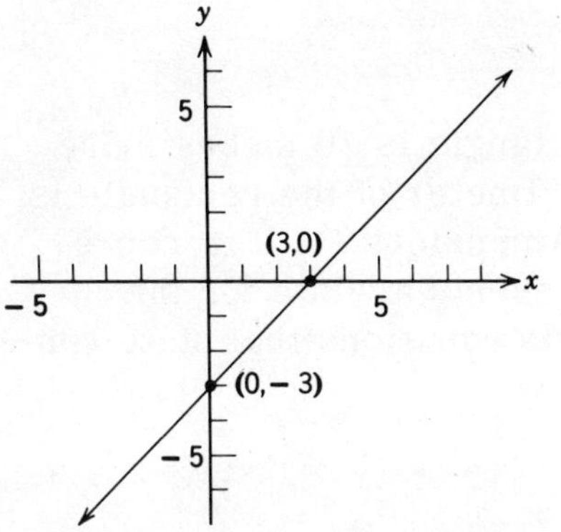

175. Locate the points corresponding to $(0,-3)$ and $(3,0)$ on the given coordinate system and complete the graph of the equation $y = x - 3$.

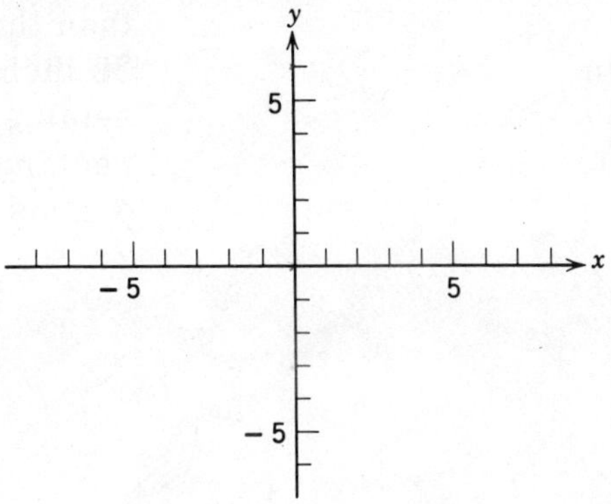

Remark. Once you have two points that lie on a line, you can easily draw the line by connecting the points. You can find the coordinates of two points that lie on the graph of an equation by choosing any value for x and using the equation to compute y, or else choosing a value for y and using the equation to get the associated value of x. The key idea here is that you can assign any value that you wish to one of the variables, and the equation will give you a value for the other.

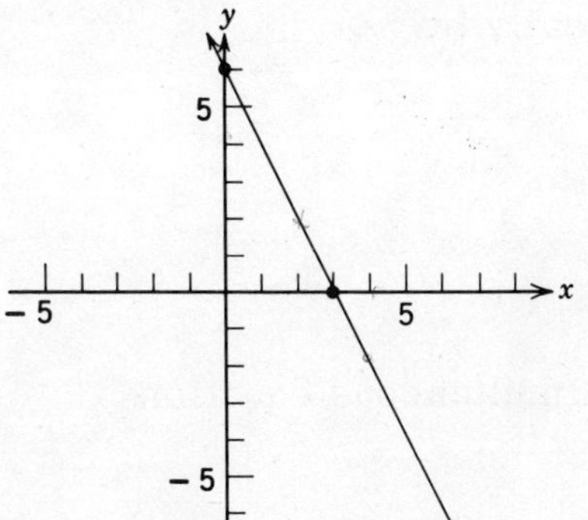

The two points you selected may have been different than ours. The graph is the same.

176. Find two solutions of the equation $y + 2x = 6$. Locate the two points whose coordinates satisfy the equation and complete the graph of $y + 2x = 6$.

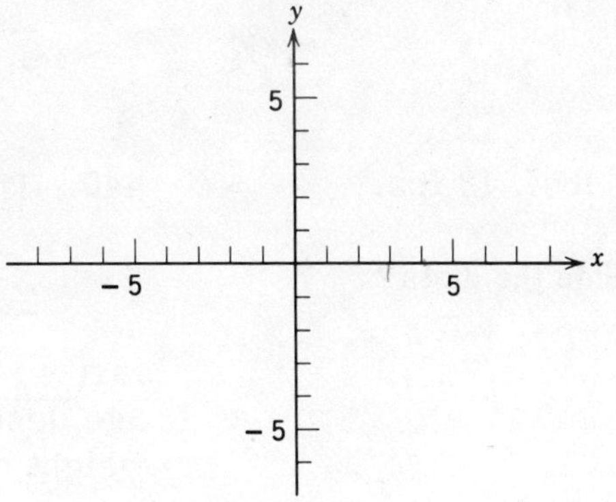

$x - y = 10$

Or $x = y + 10$
or $y = x - 10$.

437. "The length of a rectangle is 10 inches more than the width, and the perimeter of the rectangle is 28 inches. What are its dimensions?" If x represents a length and y represents a width for the rectangle, the simultaneous equations that state conditions on x and y are

$$2x + 2y = 28$$

$$\_\_\_\_\_ = \_\_\_\_\_.$$

(12, 2); 12; 2

438. The solution of the system

$$2x + 2y = 28$$
$$x - y = 10$$

is (_____, _____). Therefore, the length of the rectangle described in Frame 437 is_______ inches and the width is_____ inches.

$2x - 4 = y$

439. "Two packages weighed together total 29 pounds. one of the packages weighs 4 pounds less than twice the other. How much does each weigh?" If x and y each represent a weight, the conditions on x and y can be expressed symbolically by

$$x + y = 29$$

$$\_\_\_\_\_ = y.$$

(11, 18); 11 lbs.; 18 lbs.

Did you include the units?

440. The solution of the simultaneous equations

$$x + y = 29$$
$$2x - 4 = y$$

is (_____, _____). Therefore, the weight of one package described in Frame 439 is__________ and the weight of the other is__________.

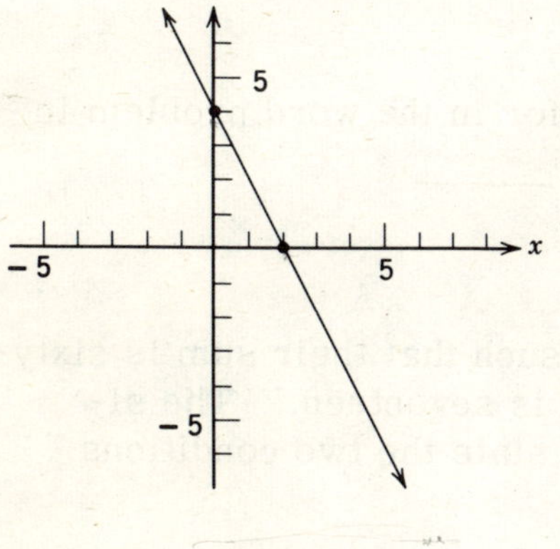

177. Graph $2x + y = 4$ by first locating two points whose coordinates satisfy the equation.

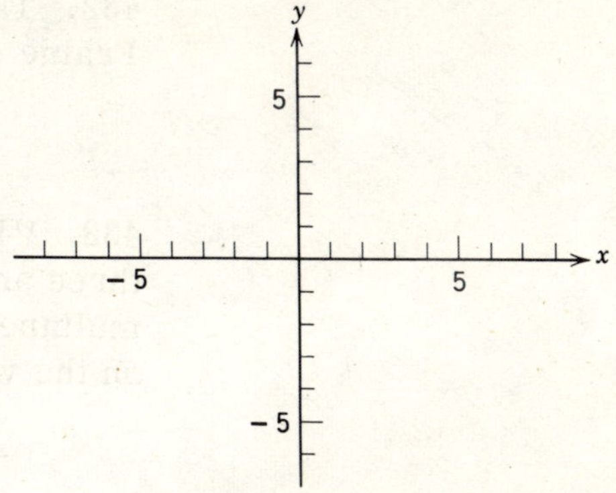

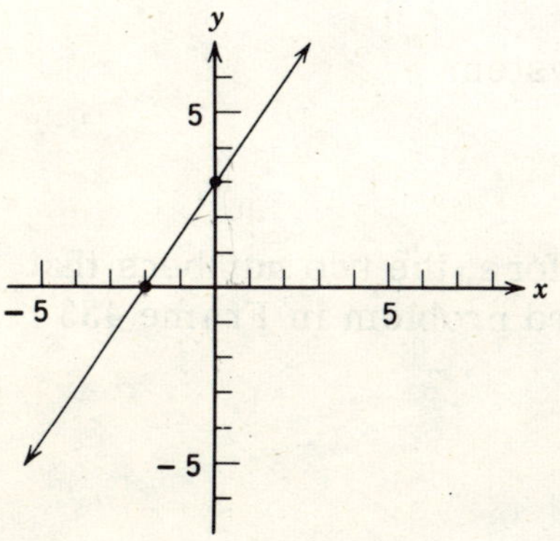

178. Graph $2y - 3x = 6$.

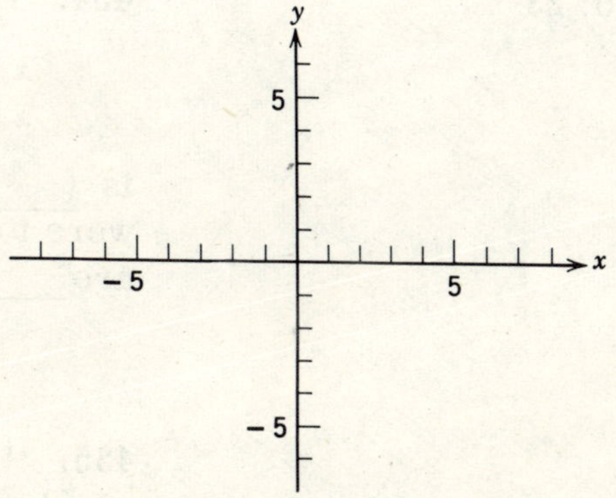

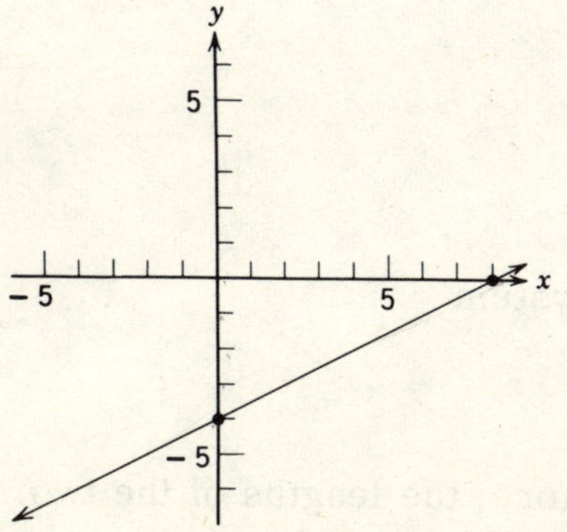

179. Graph $x - 2y = 8$.

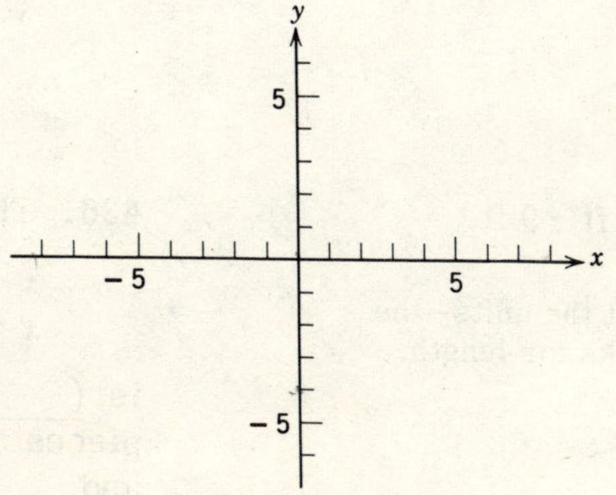

(18, 12)

431. The solution of the simultaneous equations

$$x + y = 30$$
$$x - y = 6$$

is (_____, _____).

18; 12

432. The numbers asked for in the word problem in Frame 430 are______ and______.

$x + y = 63$
$x - y = 17$

433. "Find two numbers such that their sum is sixty-three and their difference is seventeen." The simultaneous equations that state the two conditions on the variables are

________ = ________

________ = ________.

(40, 23); 40; 23

434. The solution of the system

$$x + y = 63$$
$$x - y = 17$$

is (_____, _____). Therefore, the two numbers that were to be found in the word problem in Frame 433 are______ and______.

$x + y = 20$
$x - y = 2$

435. "A 20-foot board is cut into two pieces, one of which is 2 feet longer than the other. How long is each piece?" The simultaneous equations that state the conditions on x and y are

$x + y =$_____
$x - y =$_____.

(11, 9); 11 ft.; 9 ft.

Don't forget the units—the problem asks for length.

436. The solution of the system

$$x + y = 20$$
$$x - y = 2$$

is (_____, _____). Therefore, the lengths of the two pieces of board described in Frame 435 are_______ and_______.

Remark. To find points to use in graphing an equation in two variables, you can choose any value you wish for one of the variables, and the equation will then tell you what the associated value of the other variable will be. Since you can choose any value you wish for one variable, it makes sense to choose a value that will make it easy for you to find the value of the other variable. This is what we will consider in the next sequence of frames.

(6, 0)

180. Given $x + y = 6$. Two solutions in which one component is zero are (0, 6) and (_____, 0).

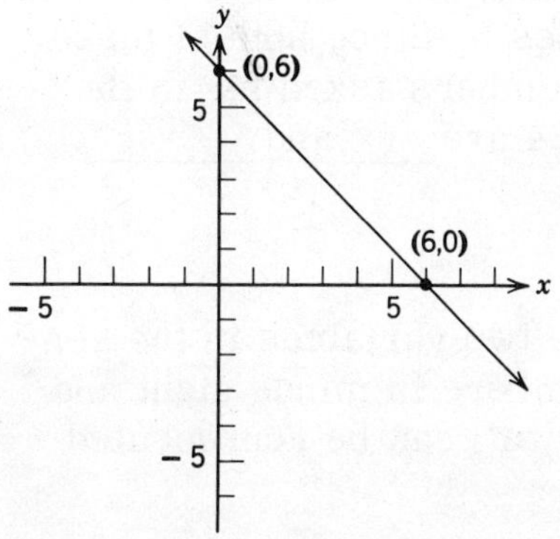

181. Graph (0, 6) and (6, 0), which are solutions of $x + y = 6$. Represent all of the solutions of the equation by drawing a straight line through these points.

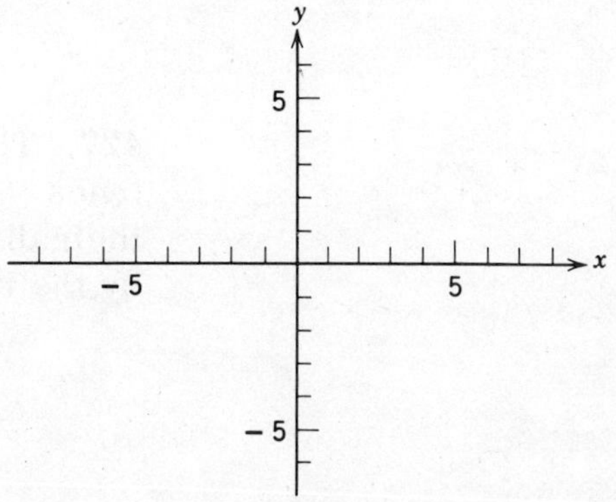

(0, 2); (4, 0)

182. Given $x + 2y = 4$. Two solutions in which one component is zero are (0, _____) and (_____, 0).

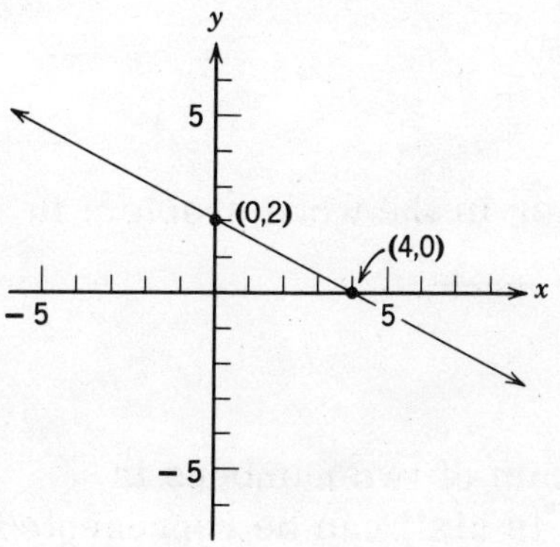

183. Graph (0, 2) and (4, 0), which are solutions of $x + 2y = 4$. Represent all of the solutions of the equation by drawing a straight line through these points.

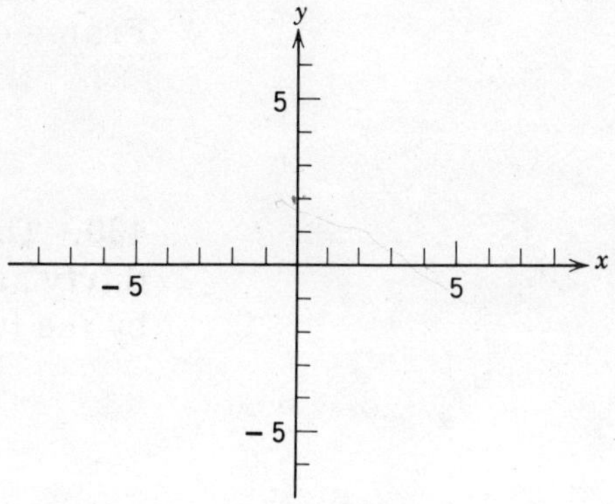

(6, 18)

425. The solution of the system of equations

$$x + y = 24$$
$$y = 3x$$

is $(\underline{\quad , \quad})$.

6; 18

426. The last step in the solution process of the word problem is to verify that the sum of 18 and 6 is 24 and that 18 is three times 6. Since both of these are true statements, the numbers asked for in the word problem in Frame 424 are____ and________.

$x + y = -8$
$x - y = -2$

427. The conditions on the two variables in the sentence "The sum of two numbers is minus eight and their difference is minus two" can be represented by the two equations

$$\underline{\qquad = \qquad}$$
$$\underline{\qquad = \qquad}.$$

(−5, −3)

428. The solution of the system

$$x + y = -8$$
$$x - y = -2$$

is $(\underline{\quad , \quad})$.

−5; −3

429. The numbers asked for in the word problem in Frame 427 are_______ and______.

$x + y = 30$
$x - y = 6$

430. The sentence "The sum of two numbers is thirty and their difference is six" can be represented by the two equations

$$\underline{\qquad = \qquad}$$
$$\underline{\qquad = \qquad}.$$

(0, −4); (−4, 0)

184. Given $x + y = -4$. Two solutions in which one component is zero are (0, ______) and (______, 0).

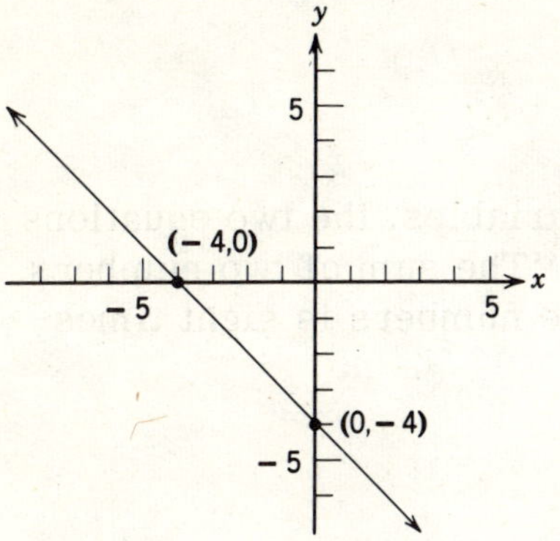

185. Graph (0, −4) and (−4, 0), which are solutions of $x + y = -4$. Represent all of the solutions of the equation by drawing a straight line through these points.

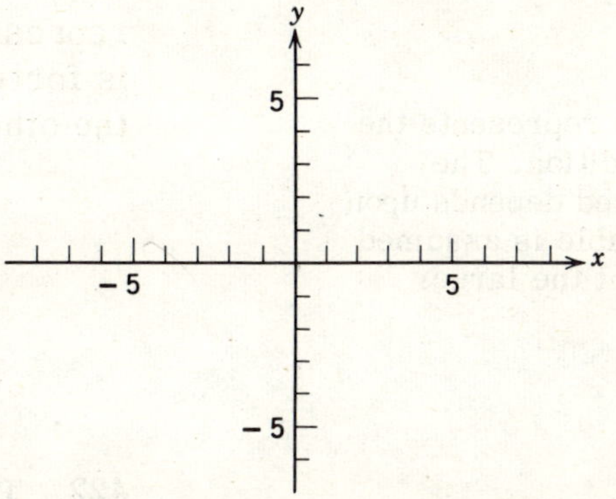

(0, −2); (2, 0)

186. Given $y = x - 2$. Two solutions in which one component is zero are (0 , −2) and (2 , 0).

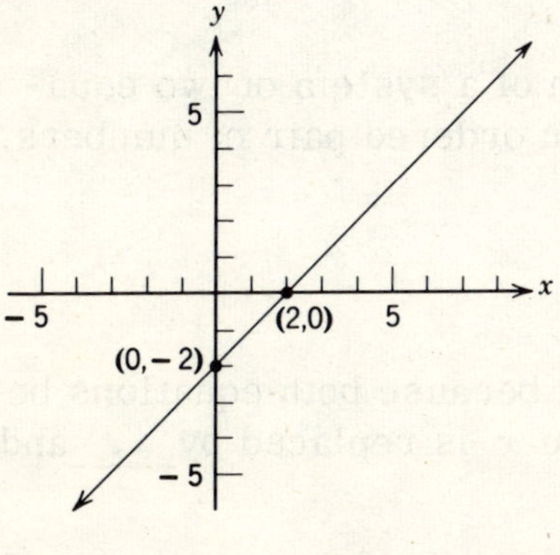

187. Graph (0, −2) and (2, 0), which are solutions of $y = x - 2$. Represent all of the solutions of the equation by drawing a straight line through these points.

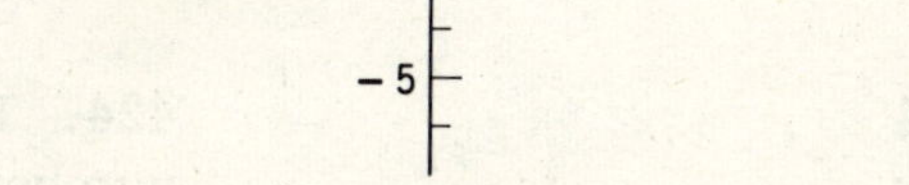

(0, 8); (−6, 0)

188. Given $3y - 4x = 24$. Two solutions in which one component is zero are (______, ______) and (______, ______).

$x + y = 25$
$x - y = 3$

420. Using x and y for variables, the sentence "The sum of two numbers is twenty-five and their difference is three" can be represented by the two equations

$$\_\_\_\_\_\_ = 25$$
$$\_\_\_\_\_\_ = 3.$$

$x + y = 45;$
$x = 8y$

$y = 8x$ also represents the second condition. The equation used depends upon which variable is assumed to represent the larger number.

421. Using x and y for variables, the two equations representing the sentence "The sum of two numbers is forty-five, and one of the numbers is eight times the other" are

$$\_\_\_\_ = \_\_\_\_$$
$$\_\_\_\_ = \_\_\_\_.$$

$x + y = 18$
$x - y = 6$

Or $y + x = 18$
$y - x = 6.$

We will not say any more about this from here on.

422. To solve the problem "Find two numbers whose sum is eighteen and whose difference is six," the conditions imposed on the numbers are first represented by the two equations

$$\_\_\_\_ = \_\_\_\_$$
$$\_\_\_\_ = \_\_\_\_.$$

12; 6

423. Recall that a solution of a system of two equations in two variables is an ordered pair of numbers. A solution of the system

$$x + y = 18$$
$$x - y = 6$$

is the ordered pair (12, 6), because both equations become true statements when x is replaced by_____ and y by____.

$x + y = 24$
$y = 3x$

424. To find a solution for the problem "Find two numbers whose sum is 24 and one of which is three times the other," the conditions imposed on the numbers are first represented by the two equations

$$\_\_\_\_ = \_\_\_\_$$
$$\_\_\_\_ = \_\_\_\_.$$

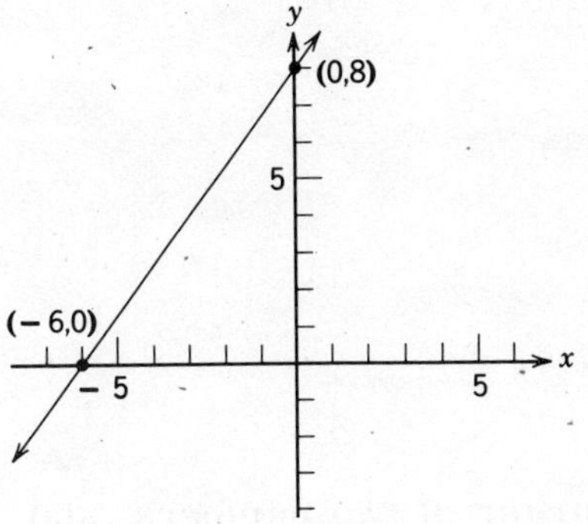

189. Graph (0, 8) and (−6, 0), which are solutions of $3y - 4x = 24$. Represent all of the solutions of the equation by drawing a straight line through these points.

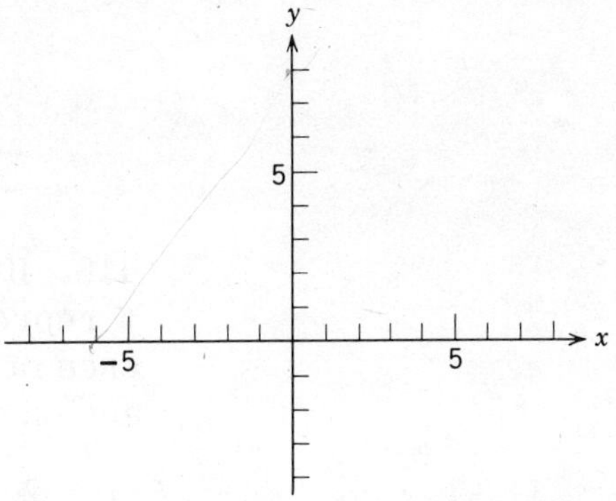

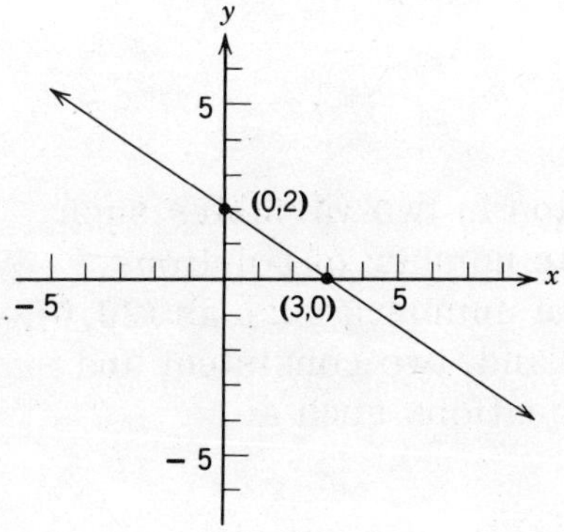

190. The solutions of $2x + 3y = 6$ that are easiest to obtain are those in which one component is zero. Find these solutions and graph the equation.

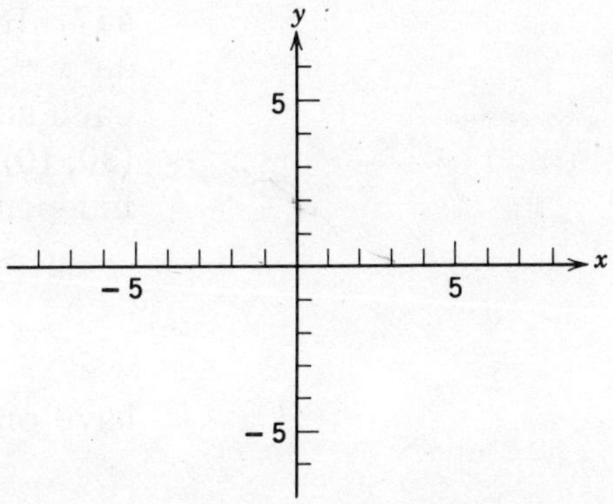

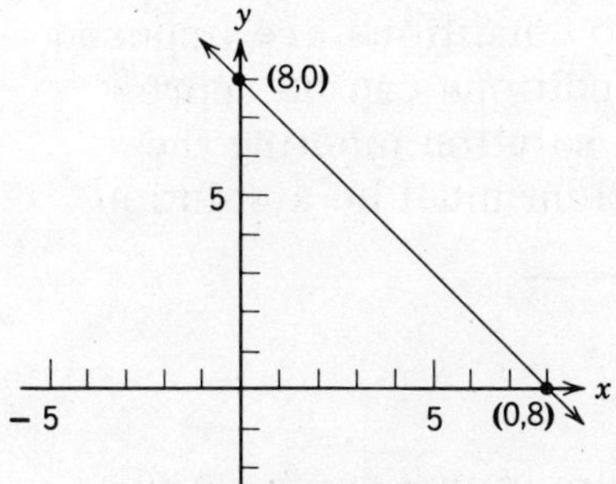

The x-intercept is 8 and the y-intercept is 8.

191. Graphing linear equations by first finding two solutions in which at least one of the components is zero is called the intercept method of graphing. Graph $x + y = 8$ by the intercept method.

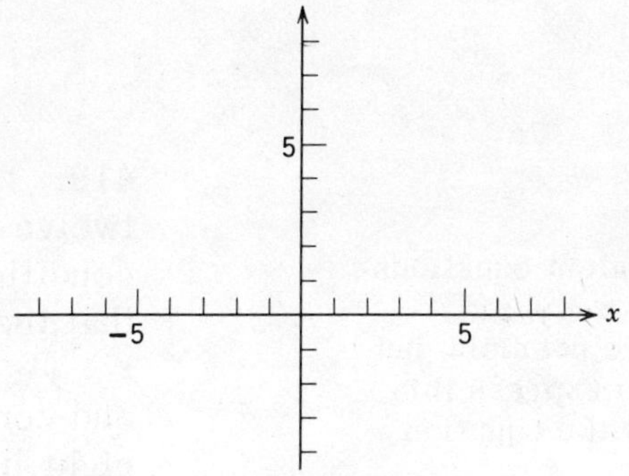

$x - y = 20$;

$x = 20 + y$;

$x - 20 = y$

416. If x represents the larger of two numbers, and y represents the smaller, the sentence "The difference of two numbers is 20" can be represented by any one of three equivalent equations:

$x - y =$ ________

$x =$ ________

________ $= y$.

one

417. Recall that one equation in two variables such as $x - y = 20$ has an infinite number of solutions, each solution being a pair of numbers such as (20, 0), (30, 10), etc. On the other hand, two consistent and independent first-degree equations such as

$$x - y = 20$$
$$x + y = 4$$

have only ________ solution in common.

equations

418. It is convenient to use two variables to solve word problems in which two conditions are imposed on the variables. These conditions can be represented by two equations. A solution meeting the conditions of the word problem must be a solution to both ________________.

$x - y = 8$

Other equivalent equations such as $x = 8 + y$ or $x - 8 = y$ are possible, but we prefer to express this condition by the equation $x - y = 8$.

419. The sentence "The sum of two numbers is twelve and their difference is eight" imposes two conditions on the two numbers. The first condition, that their sum is twelve, can be represented by $x + y = 12$, where x and y denote numbers. The second condition, that the difference of the numbers is eight, is represented by the equation ______ = ______.

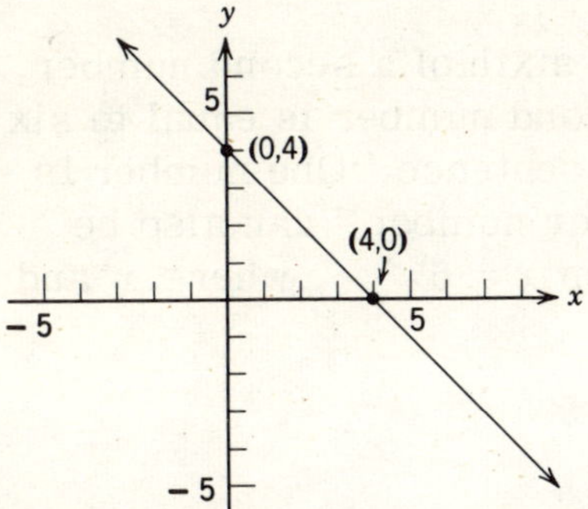

The x-intercept is 4 and the y-intercept is 4.

192. Graph $x + y = 4$ by the intercept method.

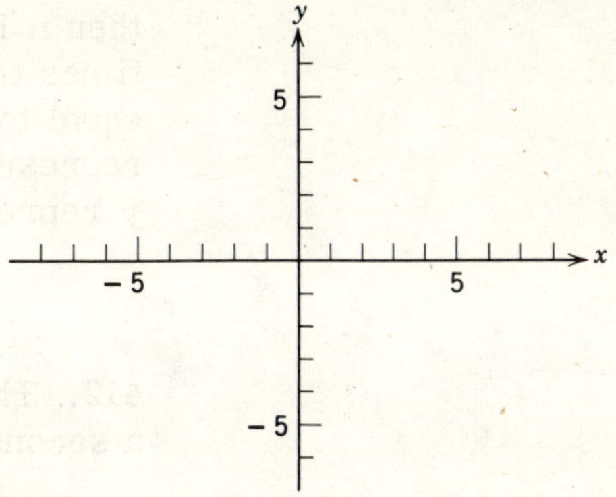

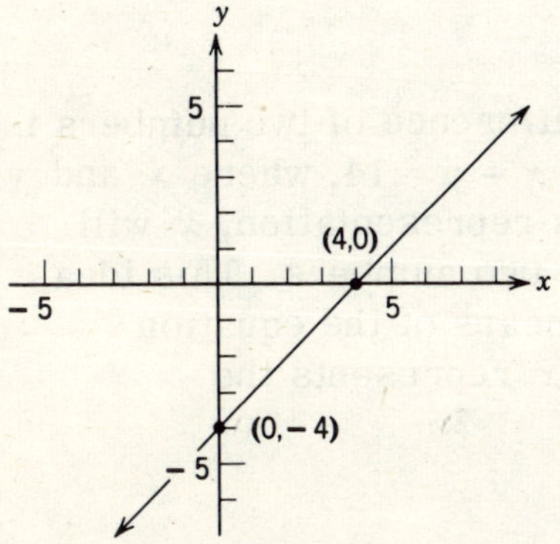

The x-intercept is 4 and the y-intercept is −4.

193. Graph $x - y = 4$ by the intercept method.

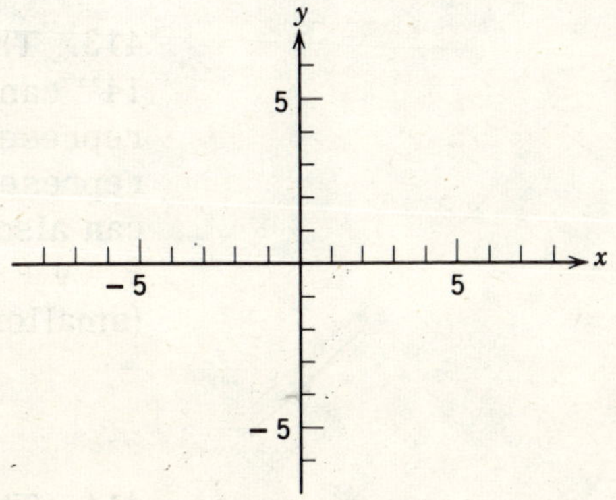

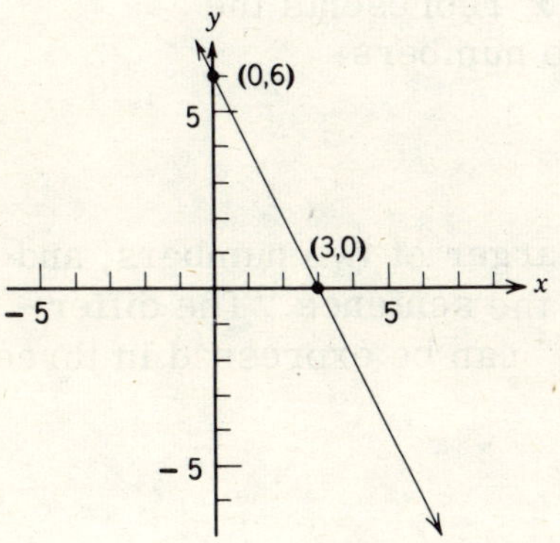

194. Graph $y = 6 - 2x$ by the intercept method.

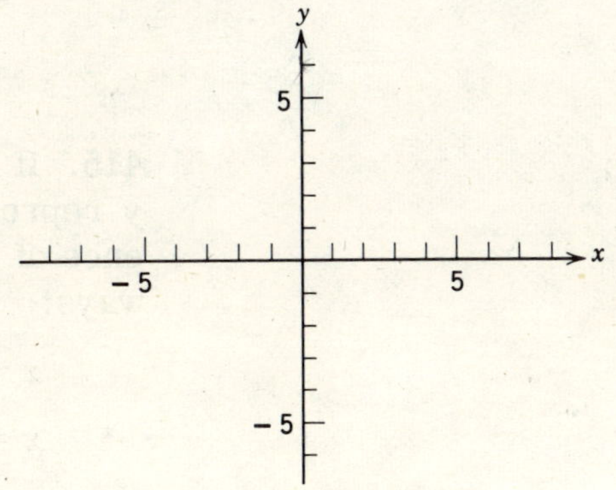

y

411. If one number is one sixth of a second number, then it follows that the second number is equal to six times the first. Thus, the sentence "One number is equal to one sixth of another number" can also be represented by the equation $x = 6$____, where x and y represent numbers.

y; x

412. The sentence "One number equals one third of a second number" can be represented by the equation ____ $= \frac{1}{3}x$ or by the equation $3y =$____, where x and y represent numbers.

larger

413. The sentence "The difference of two numbers is 14" can be represented by $x - y = 14$, where x and y represent numbers. In this representation, x will represent the larger of the two numbers. This idea can also be expressed by means of the equation $x = y + 14$, in which case, x represents the (smaller/larger) number.

larger

414. The sentence "The difference of two numbers is 14" also implies that one number is 14 less than the other. This idea can be expressed by the equation $x - 14 = y$, in which case, x represents the (smaller/larger) of the two numbers.

$x = y + 14$;

$x - 14 = y$

415. If x represents the larger of two numbers, and y represents the smaller, the sentence "The difference of two numbers is 14" can be expressed in three ways:

$x - y = 14$

$x =$________

________ $= y$.

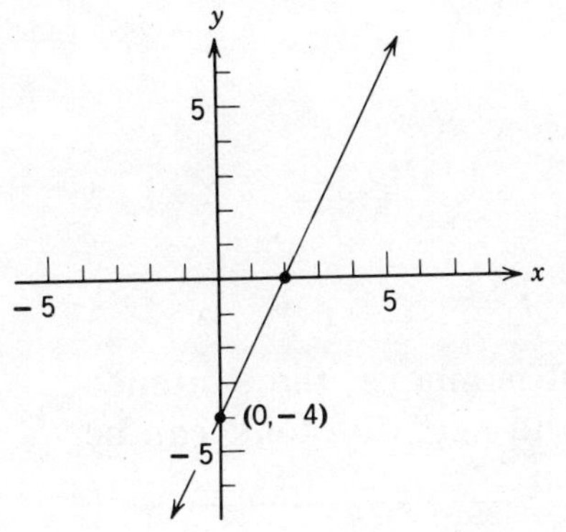

195. Graph $y = 2x - 4$.

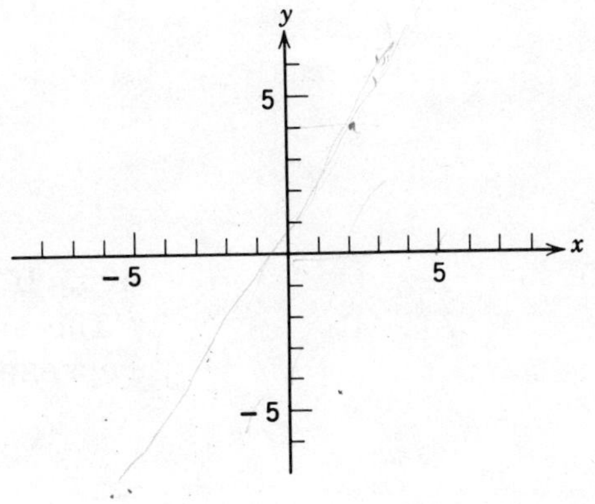

Remark. The term "intercept" comes from the fact that the graph of an equation and the two coordinate axes "intercept" each other at these points. Sometimes, however, it is not possible to use the intercept method for graphing a first-degree equation, and sometimes, while possible, it is not very satisfactory.

origin

196. Given the equation $y = 2x$. If x is 0, then y is 0, and if y is 0, then x is 0. Therefore, the graph crosses both axes at the ______________.

(3, 6)

197. One point of the graph of $y = 2x$ lies on the origin. Another point needed to determine the line representing the graph of the equation can be found by using any first or second component other than zero; one such point is designated by (3, ______).

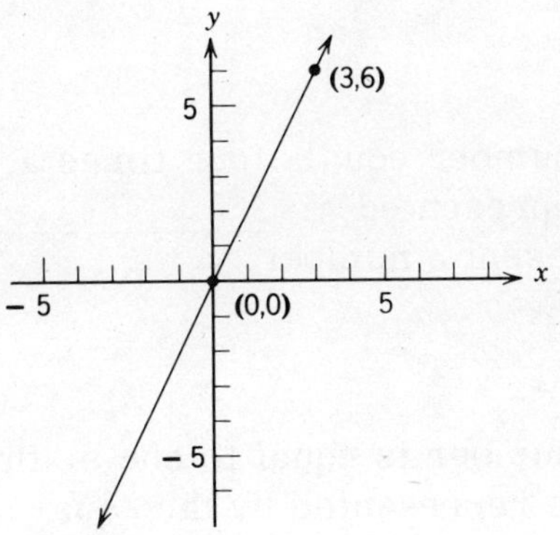

198. Two solutions of $y = 2x$ are $(0, 0)$ and $(3, 6)$. Graph the equation.

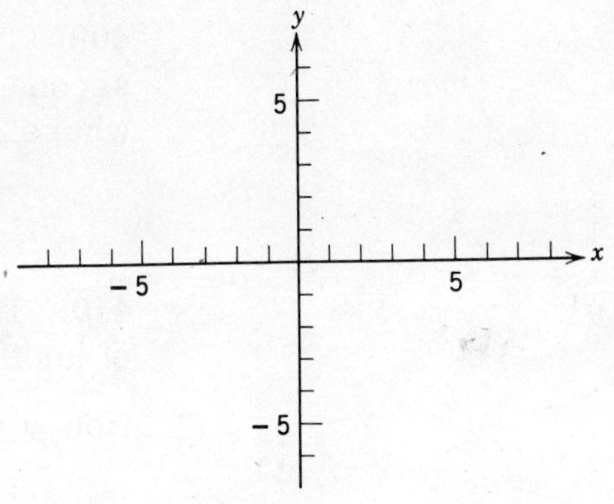

$a + b = -10$

405. If a and b represent numbers, the sentence "The sum of two numbers is negative ten" can be represented by the equation ________ = ________.

$r + s = 0$

406. Using r and s for variables, represent the sentence "The sum of two numbers is zero" by an equation.

Remark. Of course, any letters can be used for variables. However, to insure that your responses will match those in the response column, let us agree that from now on, we will use only x and y in all cases.

$x + y = -40$

407. Using x and y as variables, represent the sentence "The sum of two numbers is negative forty" by an equation.

$2y$

408. The sentence "The sum of one number and twice a second number is twenty" can be represented by $x +$ ________ $= 20$, where x and y represent numbers.

$4x$

409. The sentence "One number equals four times a second number" can be represented $y =$ ________ where x and y each represent a number.

x

410. The sentence "One number is equal to one sixth of another number" can be represented by the equation $y = \frac{1}{6}$____, where x and y represent numbers.

There are an infinite number of solutions and it would be impossible to list them. Does your ordered pair make the equation true? If it does, you have found a solution.

199. Designate any first component and determine a solution of $y = -3x$.

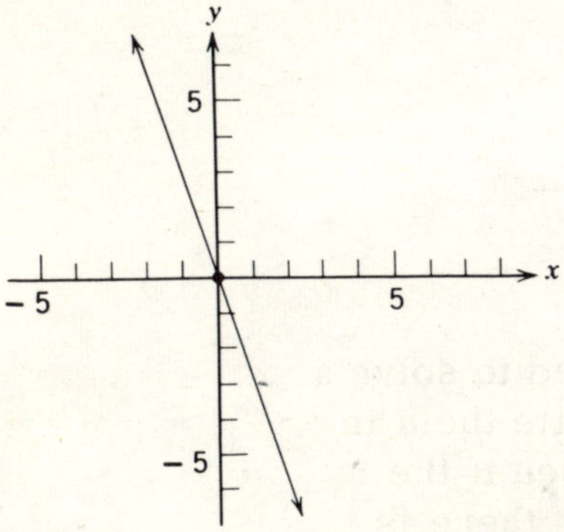

200. Graph $y = -3x$. Use $(0,0)$ and the solution you found in the previous frame.

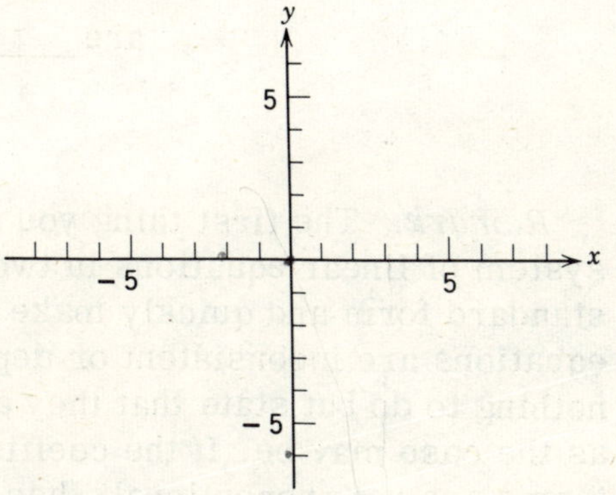

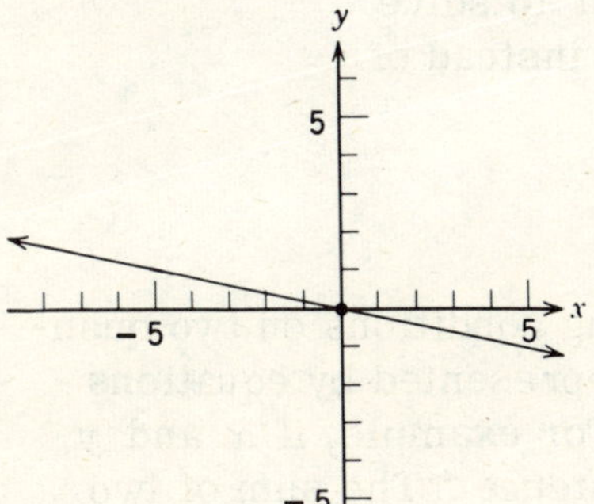

201. Graph $x + 5y = 0$. Use $(0,0)$ and any other solution you care to find to locate two points that determine the line.

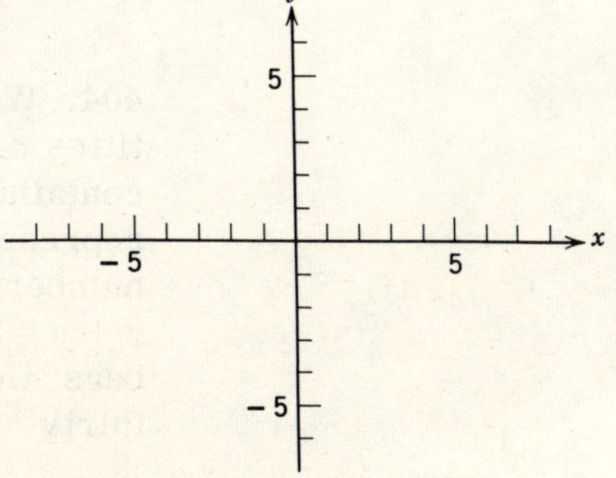

dependent

In standard form the equations appear

$x + y = 5$
$2x + 2y = 10.$

401. The equations in the system

$$x = 5 - y$$
$$2y = 10 - 2x$$

are____________________.

dependent

402. The equations in the system

$$2x = y + 6$$
$$3y = 6x - 18$$

are____________________.

inconsistent

403. The equations in the system

$$3x - y = 7$$
$$2y = 6x - 7$$

are____________________.

Remark. The first thing you should do when asked to solve a system of linear equations in two variables is to write them in standard form and quickly make a mental check to see if the equations are inconsistent or dependent. If they are, there is nothing to do but state that they are inconsistent or dependent as the case may be. If the coefficients on the like variable terms are not proportional, then you can go ahead and solve the system.

Let us next devote some attention to applications of systems of equations. It is sometimes (but not always) easier to solve a word problem by using equations in two variables instead of one equation in one variable.

$x + y = 30$

404. Word sentences stating conditions on two quantities can, in general, be represented by equations containing two variables. For example, if x and y represent numbers, the sentence "The sum of two numbers is twenty-four" can be represented by $x + y = 24$. Similarly, if x and y represent numbers, the sentence "The sum of two numbers is thirty" can be represented by the equation

__________ = __________.

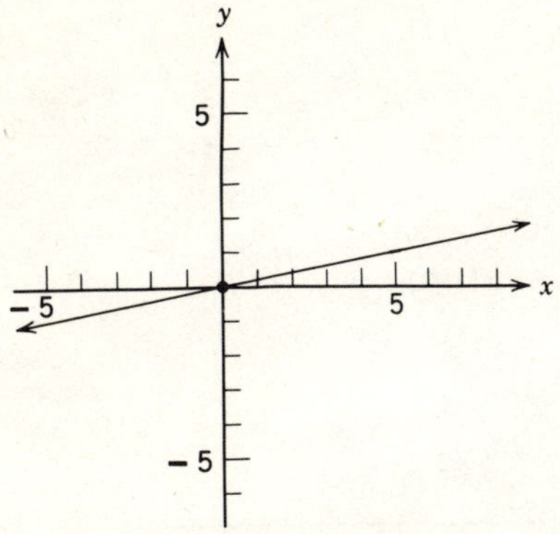

202. Graph $x - 5y = 0$.

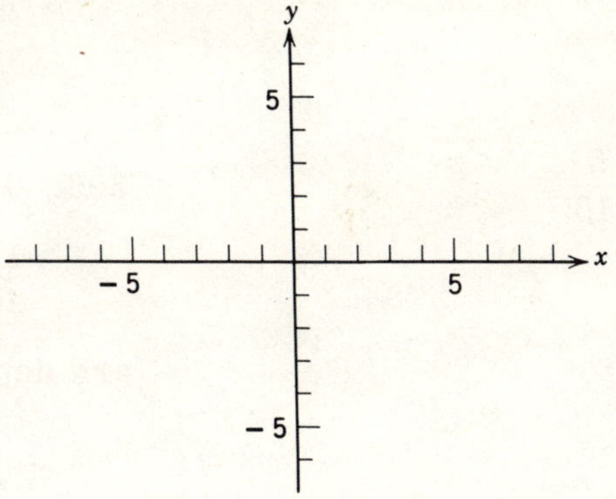

origin

203. [Figure 17] By the intercept method, two solutions of $2y = x - 1$ are $(0, -\frac{1}{2})$ and $(1, 0)$. The graph of the equation crosses both axes near the __________.

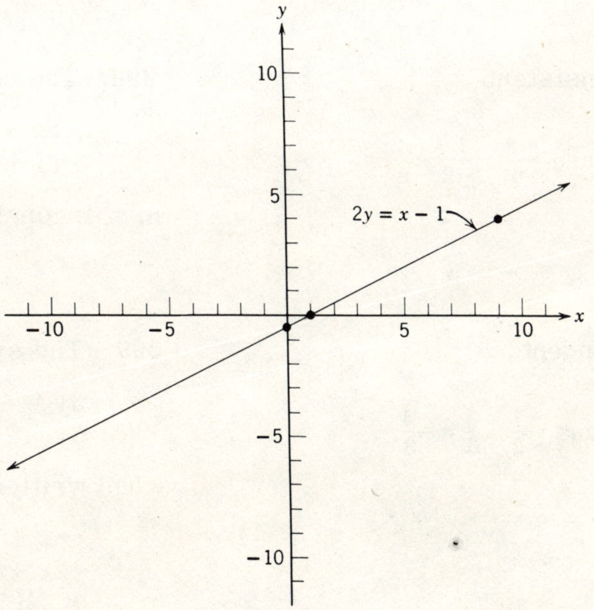

Fig. 17

graph

204. [Figure 17] It is desirable to use two solutions whose graphs are sufficiently far apart to serve as a means of accurately drawing a straight line. Therefore, a solution such as (9, 4) in conjunction with either $(0, -\frac{1}{2})$ or $(1, 0)$ would serve as a basis for drawing the____________of $2y = x - 1$.

$\frac{15}{30} = \frac{-2}{-4} = \frac{5}{10}$

396. The equations in the system

$$15x - 2y = 5$$
$$30x - 4y = 10$$

are dependent, because ______ = ______ = ______.

dependent

Because $\frac{11}{22} = \frac{1}{2} = \frac{3}{6}$.

397. The equations in the system

$$11x + y = 3$$
$$22x + 2y = 6$$

are (inconsistent/dependent).

inconsistent

Because $\frac{-3}{6} = \frac{1}{-2} \neq \frac{4}{6}$.

398. The equations in the system

$$-3x + y = 4$$
$$6x - 2y = 6$$

are (inconsistent/dependent).

dependent

Because $\frac{-1}{-2} = \frac{3}{6} = \frac{-4}{-8}$.

399. The system

$$3y = x - 4$$
$$6y = 2x - 8$$

when written in standard form appears as

$$-x + 3y = -4$$
$$-2x + 6y = -8,$$

where the equations in the system can now be seen to be (inconsistent/dependent).

inconsistent

Because $\frac{2}{6} = \frac{-1}{-3} \neq \frac{-5}{-14}$.

400. By writing the system

$$2x = y - 5$$
$$6x = 3y - 14$$

in standard form, the equations in the system can be seen to be (inconsistent/dependent).

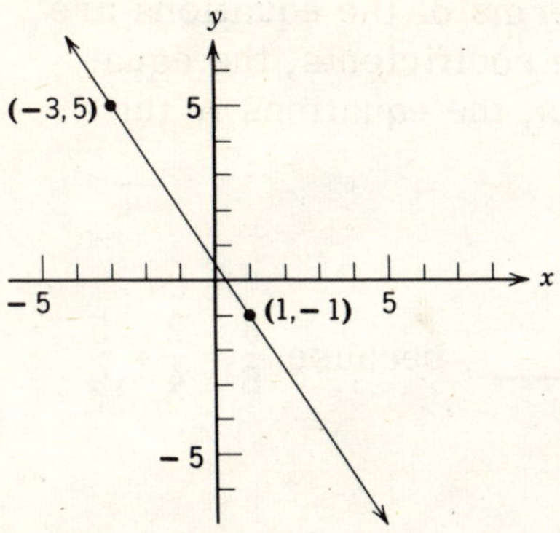

205. Two solutions of $3x + 2y = 1$ are $(1, -1)$ and $(-3, 5)$. Graph the equation.

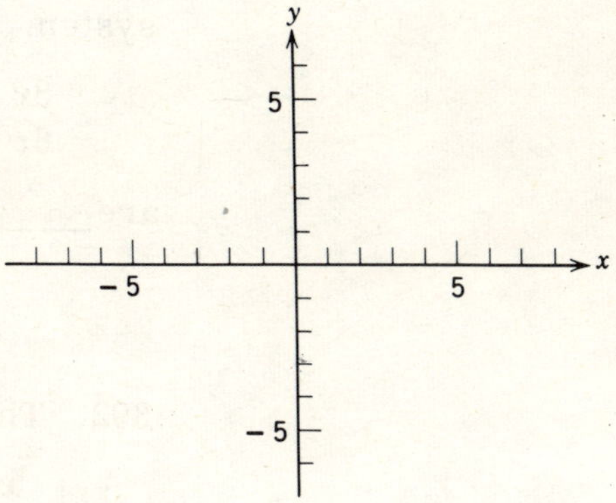

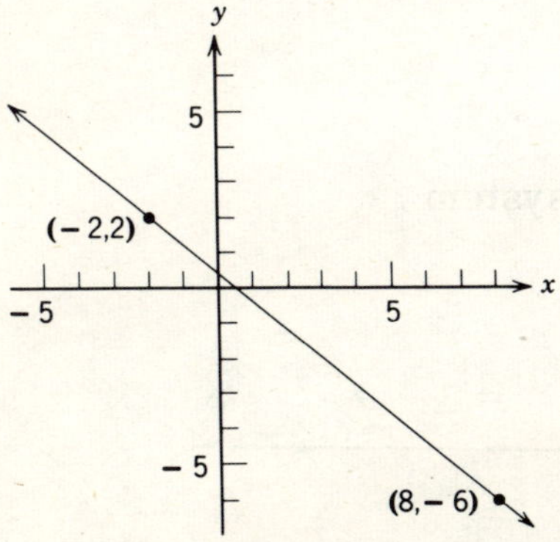

206. Graph $5y = 2 - 4x$. Use the solutions $(-2, 2)$ and $(8, -6)$.

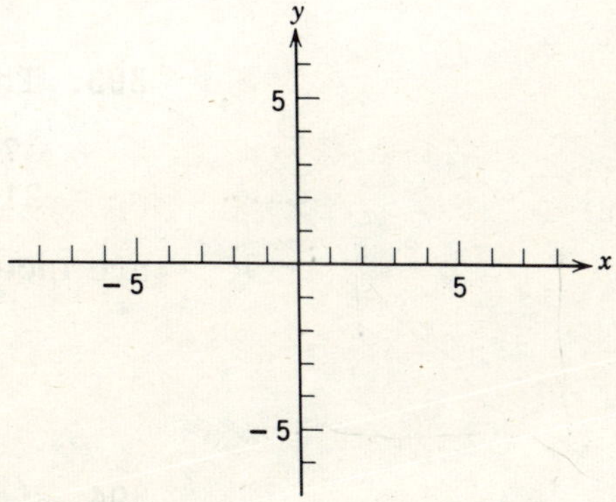

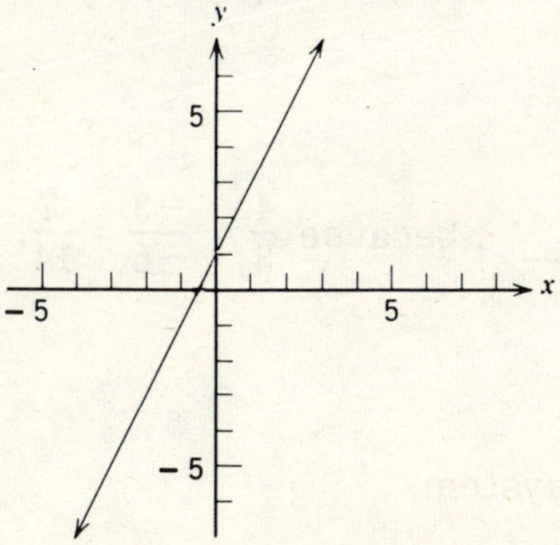

207. Graph $y = 1 + 2x$. Use two solutions whose graphs are sufficiently far apart.

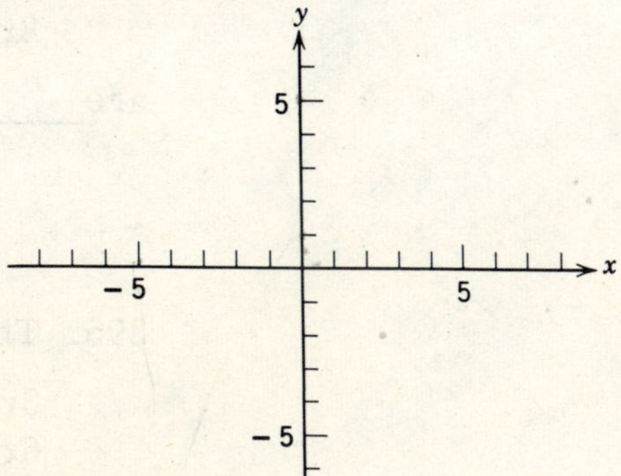

inconsistent

391. If the coefficients of the like variables of the equations in a system in standard form are proportional, while the constant terms of the equations are not in the same ratio as the coefficients, the equations are inconsistent. Thus, the equations in the system

$$3x + 2y = 5$$
$$6x + 4y = 9$$

are ____________________, because $\frac{3}{6} = \frac{2}{4} \neq \frac{5}{9}$.

$\frac{7}{15}$

392. The equations in the system

$$5x - 3y = 7$$
$$10x - 6y = 15$$

are inconsistent, because $\frac{5}{10} = \frac{-3}{-6} \neq$ ______.

$\frac{7}{21} = \frac{2}{6} \neq \frac{5}{11}$

393. The equations in the system

$$7x + 2y = 5$$
$$21x + 6y = 11$$

are inconsistent, because ___ = ___ ≠ ___.

dependent

394. If the constant terms of the equations in a system in standard form are in the same ratio as the coefficients of the like variables in the equations, the equations are dependent. Thus, the equations in the system

$$4x - 3y = 7$$
$$8x - 6y = 14$$

are ____________________, because $\frac{4}{8} = \frac{-3}{-6} = \frac{7}{14}$.

$\frac{7}{14}$

395. The equations in the system

$$3x - 2y = 7$$
$$6x - 4y = 14$$

are dependent, because $\frac{3}{6} = \frac{-2}{-4} =$ ______.

Remark. By now, you can graph any first-degree equation in two variables. Such graphs are useful for a variety of things. You can tell quite a bit about the relationship that exists between the variables involved by simply inspecting the graph. For example, if the graph is rising from left to right, then the value of y is increasing as the value of x increases, while if the graph is falling from left to right, then the value of y is decreasing while the value of x increases. Another use to which we can put graphs is to find common solutions, if there are any, for two equations in two variables.

infinite

208. The equation $x + y = 7$ has an infinite number of solutions. Some of these solutions are (0, 7), (1, 6), (2, 5), (3, 4), (4, 3), etc. Any equation in two variables has an ______________ number of solutions.

infinite

209. The equation $3y - x = 9$ has an ______________ number of solutions.

solution

210. [Figure 18] The graph of $3y - x = 9$ is the straight line as shown. The line consists of an infinite number of points each of which has as coordinates an ordered pair that is a ______________ of the equation $3y - x = 9$.

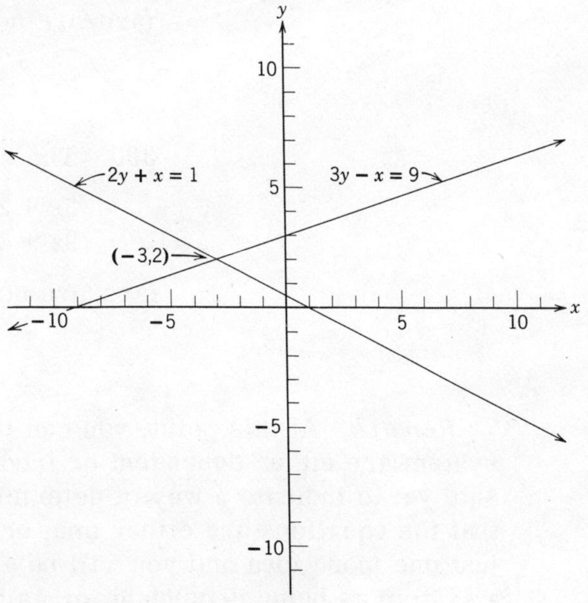

Fig. 18

inconsistent; dependent

Or dependent; inconsistent.

387. The coefficients of the like variables of the equations in the system

$$5x + y = 7$$
$$15x + 3y = 21$$

form the proportion $\frac{5}{15} = \frac{1}{3}$. Therefore, the equations of the system are either ____________________ or ____________________.

are

$\frac{5}{-5}$ does equal $\frac{-3}{3}$.

388. The coefficients of the like variables of the equations in the system

$$5x - 3y = 7$$
$$-5x + 3y = 2$$

(are/are not) proportional.

are not

389. The equations in the system

$$5x - 3y = 7$$
$$-5x - 3y = 2$$

(are/are not) either dependent or inconsistent.

are

390. The equations in the system

$$3x + 2y = 7$$
$$9x + 6y = 18$$

(are/are not) either dependent or inconsistent.

Remark. At this point, you can tell whether the equations in a system are either dependent or inconsistent, but nothing has been said yet to indicate a way of determining which. All you can say is that the equations are either one, or the other, or neither. However, just one more idea and you will be able to identify the equations in a system as being dependent, or as being inconsistent, or as being neither, which is to say, as being consistent and independent.

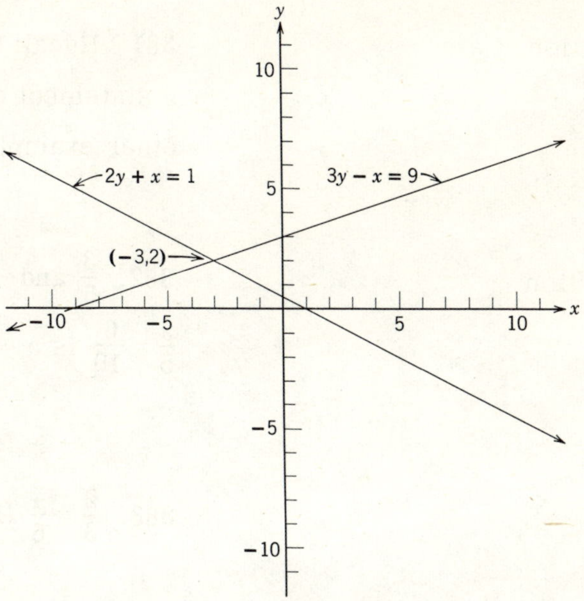

Fig. 18

solution

211. [Figure 18] The graph of $2y + x = 1$ is the straight line as shown. The line consists of an infinite number of points each of which has as coordinates an ordered pair that is a ________________ of the equation $2y + x = 1$.

point

212. [Figure 18] The point that lies on both lines is called the point of intersection of the lines. Any ordered pair that is a solution of both $3y - x = 9$ and $2y + x = 1$ must have components that are the coordinates of a ____________ that lies on both lines.

solution

213. [Figure 18] The point of intersection as shown on the graph has coordinates -3 and 2. The ordered pair $(-3, 2)$ is a solution of $3y - x = 9$ because $3(2) - (-3)$ equals 9. $(-3,2)$ is also a ____________ of $2y + x = 1$ because $2(2) + (-3) = 1$.

simultaneous

214. Two equations considered together are called a system of simultaneous equations. $3y - x = 9$ and $2y + x = 1$ when considered together form a system of ________________________ equations.

proportion

381. Recall that a proportion, for example $\frac{2}{4} = \frac{3}{6}$, is a statement equating two equal ratios. $\frac{1}{3} = \frac{2}{6}$ is another example of a ______________________.

proportion

382. $\frac{3}{5}$ and $\frac{6}{10}$ are equal ratios, therefore, the statement $\frac{3}{5} = \frac{6}{10}$ is a ______________________.

is

383. $\frac{2}{3} = \frac{4}{6}$ (is/is not) a proportion.

is not

384. $\frac{5}{10} = \frac{2}{3}$ (is/is not) a proportion.

$\frac{5}{10}$ does not equal $\frac{2}{3}$.

is

385. Recall that a system of equations is in standard form if the corresponding terms in the left members of each equation contain the same variables, and the right members of both equations do not contain variables. Thus, the system

$$\begin{aligned} x + y &= 7 \\ 2x + 2y &= 9 \end{aligned}$$

(is/is not) in standard form.

proportion

386. In a system that is written in standard form, if the coefficients of the same variables are proportional, the equations in the system are either dependent or inconsistent. Thus, the equations in the system

$$\begin{aligned} 2x + 4y &= 7 \\ 4x + 8y &= 15 \end{aligned}$$

are either dependent or inconsistent, because the coefficients of x and y form the ______________________ $\frac{2}{4} = \frac{4}{8}$.

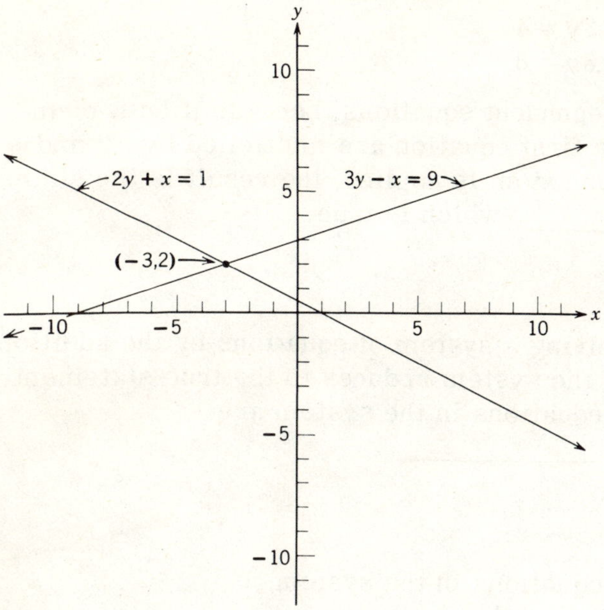

Fig. 18

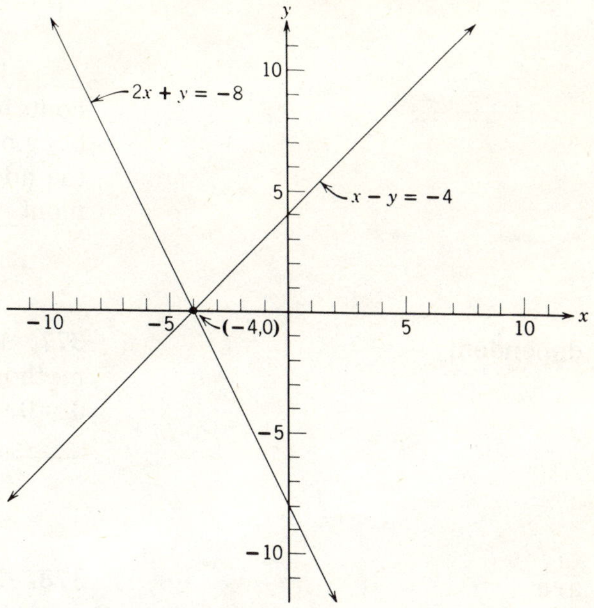

Fig. 19

simultaneous

215. The two equations

$$2x + y = 6$$
$$x - y = 3$$

when considered together are called a system of ______________ equations.

intersection

216. If a system of two simultaneous equations in two variables has a single solution, the solution is the ordered pair of numbers associated with the point of ______________ of their graphs.

(−3, 2)

217. [Figure 18] The solution of the system

$$3y - x = 9$$
$$2y + x = 1$$

is (______ , ______).

(−4, 0)

218. [Figure 19] The solution of the system

$$2x + y = -8$$
$$x - y = -4$$

is (______ , ______).

$0 = 0$

376. The system

$$x - 3y = 4$$
$$2x - 6y = 8$$

contains dependent equations, because if both members of the first equation are multiplied by -2 and the addition axiom is applied, the result is the statement ______=______ which is true.

dependent

377. In solving a system of equations by the addition method, if the system reduces to the true statement $0 = 0$, the equations in the system are ____________________.

are

378. The equations in the system

$$3x + y = 4$$
$$6x + 2y = 8$$

(are/are not) dependent.

inconsistent

Multiplying the first equation by -2 and adding the left and right members of the equations results in the false statement $0 = 1$.

379. The equations in the system

$$3y - 4x = 7$$
$$6y - 8x = 15$$

are (inconsistent/dependent).

dependent

Multiplying the first equation by -2 and adding the left and right members of the equations results in the statement $0 = 0$ which is true.

380. The equations in the system

$$5y - 3x = 8$$
$$10y - 6x = 16$$

are (inconsistent/dependent).

Remark. It is possible to recognize inconsistent and dependent equations by simply inspecting the coefficients of the variables and the constant terms. This is the idea brought out in the next sequence of frames, following a brief review of the notion of ratio and proportion.

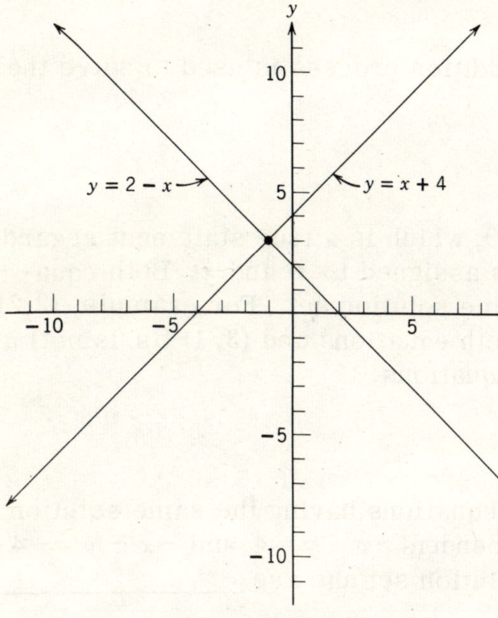

Fig. 20

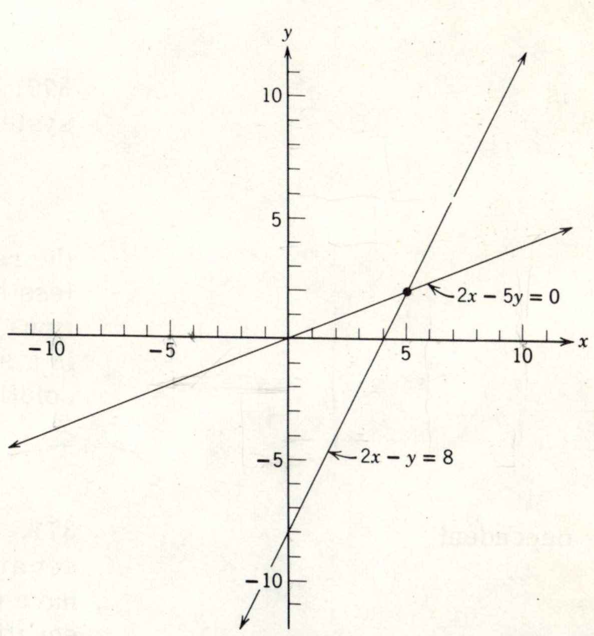

Fig. 21

(−1, 3)

Since you estimated the coordinates, perhaps you did not obtain exactly the same numbers we did.

219. [Figure 20] Estimate the coordinates of the point of intersection of the graphs of the simultaneous equations

$$y = x + 4$$
$$y = 2 - x.$$

solution

220. [Figure 20] (−1, 3) is a solution of $y = x + 4$, because $(3) = (-1) + 4$. (−1, 3) is a solution of $y = 2 - x$ also, because $(3) = 2 - (-1)$. Therefore, (−1, 3) is said to be the ________ of the system

$$y = x + 4$$
$$y = 2 - x.$$

(5, 2)

You can check this by replacing x with 5 and y with 2 in each equation.

221. [Figure 21] The solution of the system

$$2x - 5y = 0$$
$$2x - y = 8$$

is (___ , ___).

is

370. When the addition process is used to solve the system

$$x + y = 4$$
$$-x - y = -4,$$

the result is $0 = 0$, which is a true statement regardless of the values assigned to x and y. Both equations have the same solution set. For example, (2,2) is a solution of both equations and (3, 1) (is/is not) a solution of both equations.

dependent

371. Recall that equations having the same solution set are called dependent. $x + y = 4$ and $-x - y = -4$ have the same solution set and are ____________ equations.

dependent

372. If the result of adding the respective members of the equations in a system is a statement containing no variables, and if the resulting statement is a true statement, the equations in the system have identical solution sets and are ____________.

$0 = 0$

373. The result of applying the addition axiom to the system

$$3x - y = 4$$
$$-3x + y = -4$$

is the statement ____ = ____.

true

374. The statement $0 = 0$ is (true/false).

dependent

375. The equations $3x - y = 4$ and $-3x + y = -4$ have identical solution sets and are ____________.

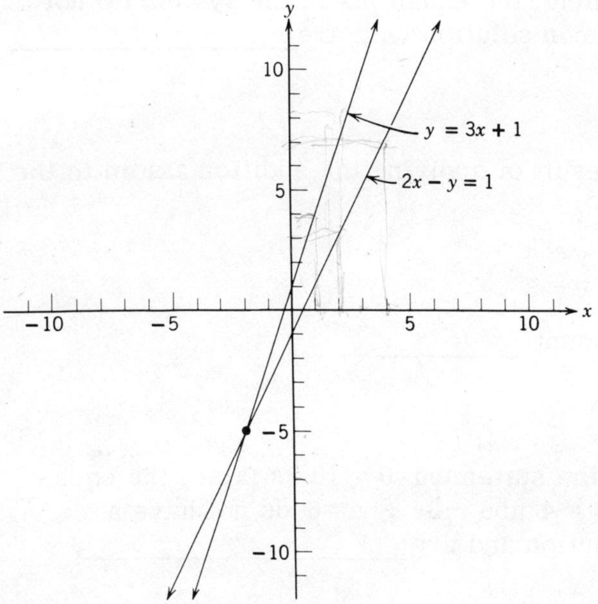

Fig. 22

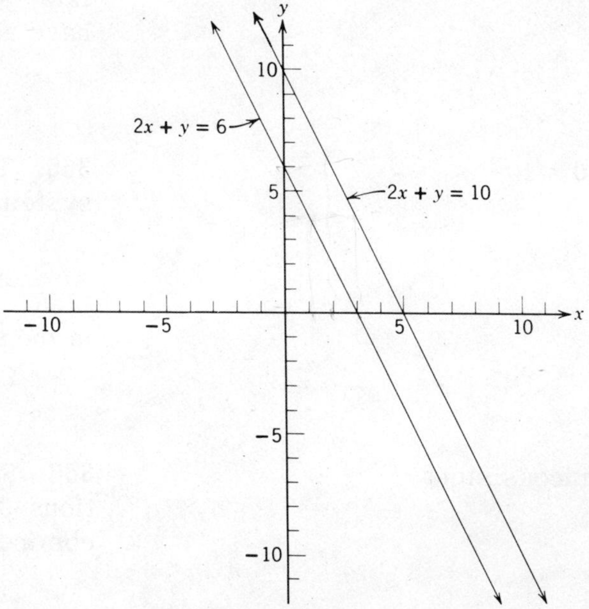

Fig. 23

(−2, −5)

222. [Figure 22] The solution of the system

$$y = 3x + 1$$
$$2x - y = 1$$

is (________, ________).

inconsistent

223. [Figure 23] Two equations whose graphs are parallel lines do not have an ordered pair common to the solution set of each. Such equations are called inconsistent. From an inspection of their graphs,

$$2x + y = 6$$
$$2x + y = 10$$

appear to be________________________ equations.

inconsistent

364. If the result of adding the respective members of the equations in a system is a statement containing no variables, and if the resulting statement is a false statement, the equations in the system do not have a common solution, and are ____________________.

$0 = 10$

365. The result of applying the addition axiom to the system

$$\begin{aligned} 3x - y &= 4 \\ -3x + y &= 6 \end{aligned}$$

is the statement ______ = ______.

inconsistent

366. Since the statement $0 = 10$ is false, the equations $3x - y = 4$ and $-3x + y = 6$ do not have a common solution and are ____________________.

$0 = 3$

367. The system

$$\begin{aligned} x + 3y &= 4 \\ 2x - 6y &= 11 \end{aligned}$$

contains inconsistent equations. If both members of the first equation are multiplied by -2, and the addition axiom is then applied, the result is the statement ______ = ______, which is false.

inconsistent

368. In solving a system of equations by the addition method, if the system reduces to a false statement, the equations in the system are ____________________.

are

369. The equations in the system

$$\begin{aligned} 3x + y &= 4 \\ 6x + 2y &= 11 \end{aligned}$$

(are/are not) inconsistent.

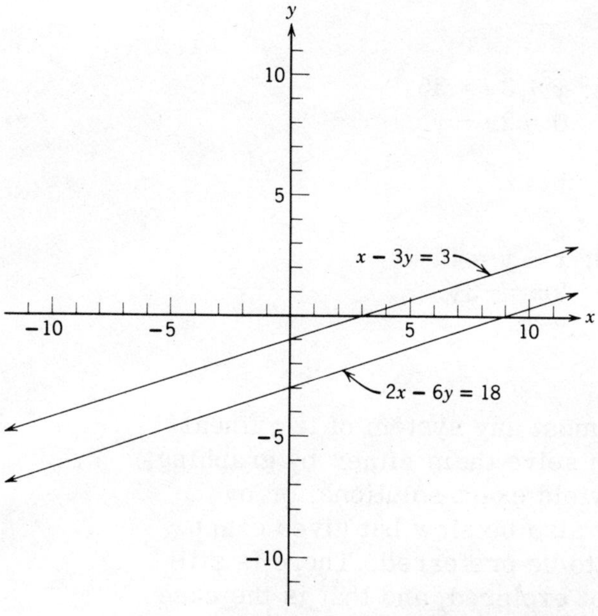

Fig. 24

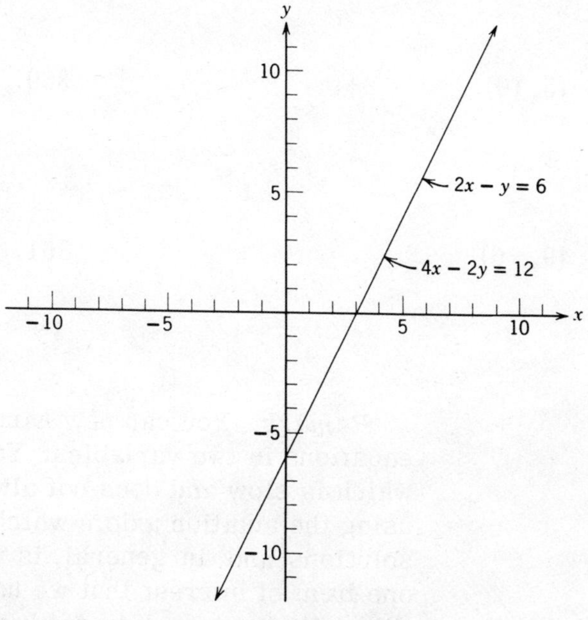

Fig. 25

inconsistent

224. [Figure 24] From an inspection of their graphs,

$$x - 3y = 3$$
$$2x - 6y = 18$$

appear to be____________________equations.

dependent

Dependent equations are equivalent equations.

225. [Figure 25] Two equations whose graphs are the same line have an infinite number of ordered pairs common to the solution set of each. Such equations are called dependent.

$$2x - y = 6$$
$$4x - 2y = 12$$

represent____________________equations.

(1, 1)

358. Solve: $7 = 4x + 3y$
$x = 5y - 4.$

(1, 2)

359. Solve: $3x + 2y = 7$
$x = y - 1.$

(5, 10)

360. Solve: $x + 3y = 35$
$0 = 2x - y.$

(9, −6)

361. Solve: $x - y = 15$
$2x = -3y.$

Remark. You can now handle almost any system of two linear equations in two variables. You can solve them either by graphing, which is slow and does not always yield exact solutions, or by using the addition axiom which may also be slow but gives exact solutions and, in general, is much to be preferred. There is still one item of interest that we have not explored, and that is the case where the system does not have just a single solution. Sometimes a system is dependent, and has an infinite number of solutions, and sometimes a system is inconsistent and does not have a solution at all. How does this show up when we are trying to solve the system by applying the addition axiom?

does not

362. When the addition axiom is applied to solve the system

$$x + 2y = 4$$
$$-x - 2y = 8,$$

the process leads to the statement $0 = 12$. Since there are no values of x or y for which this statement is true, the system (does/does not) have a solution.

inconsistent

363. Recall that the equations in systems that have no solution are called inconsistent. Since $x + 2y = 4$ and $-x - 2y$ do not have a common solution, they are ________________ equations.

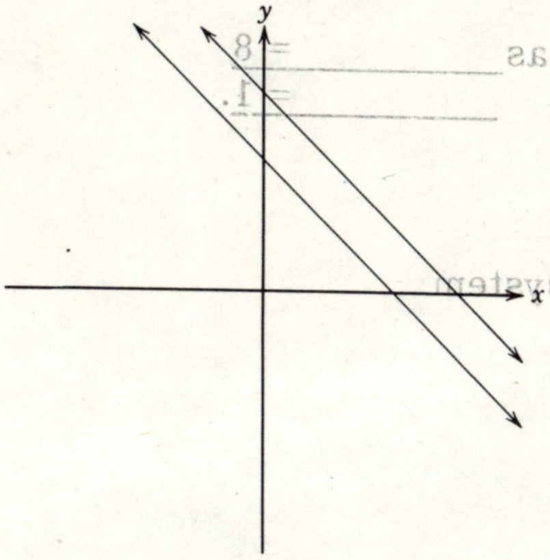

Fig. 26

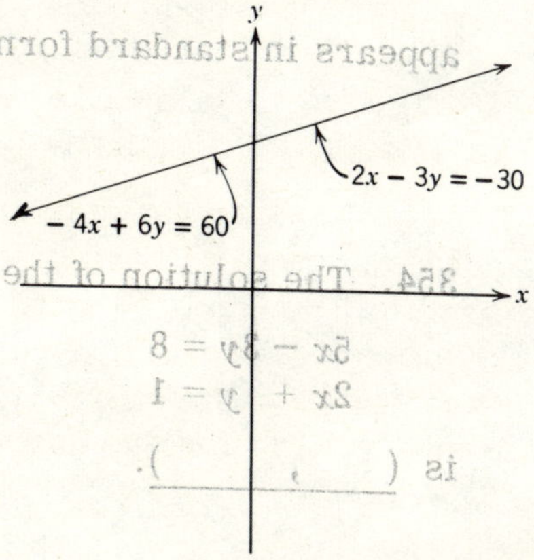

Fig. 27

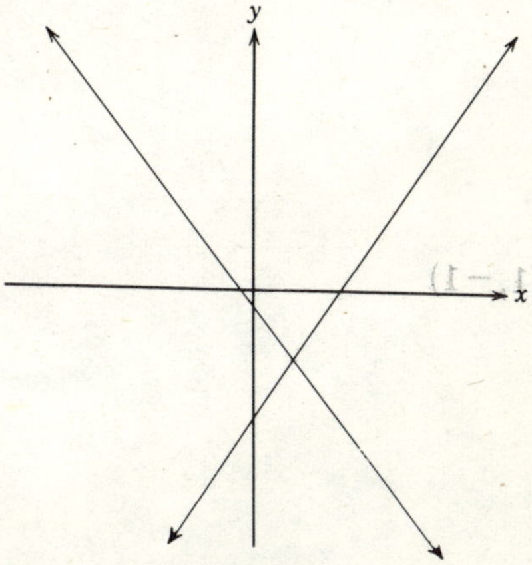

Fig. 28

inconsistent

226. [Figure 26] The figure appears to show the graphs of a system of ______________________ equations.

dependent

There is only one graph for both equations.

227. [Figure 27] The figure appears to show the graph of a system of ________________ equations.

solution

228. [Figure 28] The figure shows the graphs of two equations that have a single ________________ in common.

$(-3, 2)$

352. The solution of the system

$$2x + 3y = 0$$
$$5x - 2y = -19$$

is (____, ____).

$5x - 3y = 8$
$2x + y = 1$

353. The system

$$5x = 3y + 8$$
$$2x = 1 - y$$

appears in standard form as ________ $= 8$
________ $= 1$.

$(1, -1)$

354. The solution of the system

$$5x - 3y = 8$$
$$2x + y = 1$$

is (____, ____).

$x + 10y = 0$
$3x + y = 29$

355. The system

$$x = -10y$$
$$3x = 29 - y$$

appears in standard form as ________ $= 0$
________ $= 29$.

$(10, -1)$

356. The solution of the system

$$x + 10y = 0$$
$$3x + y = 29$$

is (____, ____).

$(5, 7)$

357. The solution of the system

$$5x - 3y = 4$$
$$y = x + 2$$

is (____, ____).

(7, 3)

Perhaps your estimate was a little different from ours.

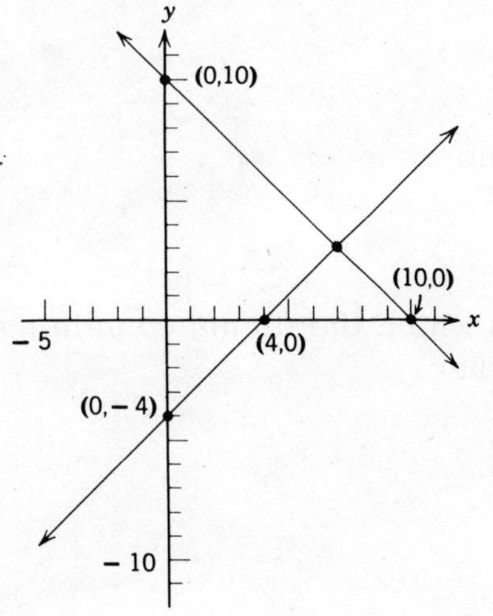

229. Two solutions of $x + y = 10$ are $(0, 10)$ and $(10, 0)$ and the graph of this equation is shown. Two solutions of $x - y = 4$ are $(0, -4)$ and $(4, 0)$ and these two solutions are also graphed. Complete the graphs of $x - y = 4$, and estimate the coordinates of the point of intersection.

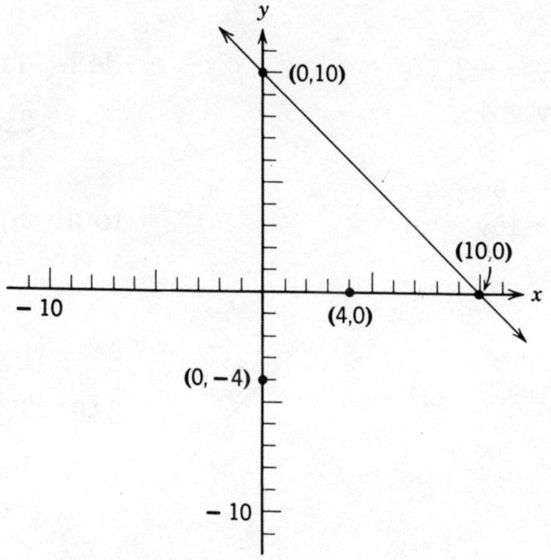

(7, 3)

230. Since 7 and 3 are the coordinates of the point of intersection of the graphs of $x + y = 10$ and $x - y = 4$, the solution of the system is (____, ____).

(4, –4)

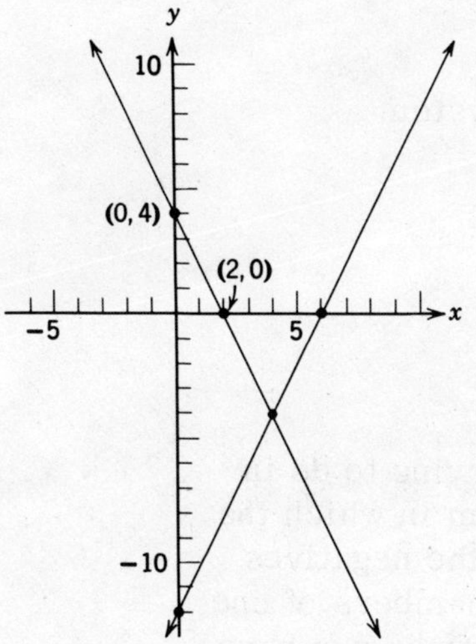

231. The graph of $2x - y = 12$ is shown. Two solutions of $2x + y = 4$ are $(0, 4)$ and $(2, 0)$; these two solutions are also graphed. Complete the graph of $2x + y = 4$ and estimate the coordinates of the point of intersection.

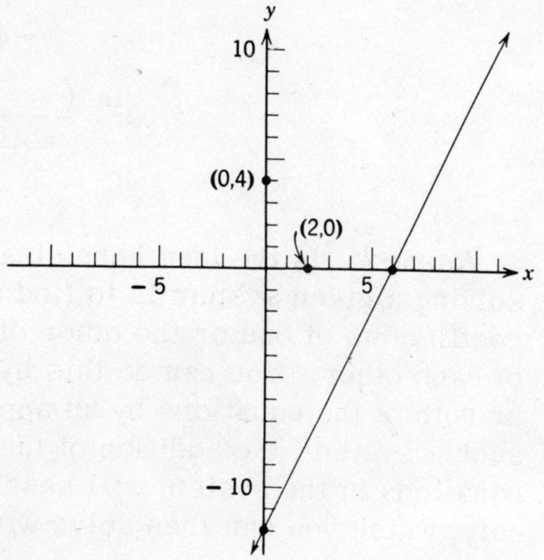

$6x + 9y = -3$
$-6x - 10y = 4$

Or $-6x - 9y = 3$
$6x + 10y = -4$.

348. Transform the system

$$2x + 3y = -1$$
$$3x + 5y = -2$$

to an equivalent system in which the terms containing x are negatives of each other.

(1,−1)

349. The solution of

$$6x + 9y = -3$$
$$-6x - 10y = 4$$

is (_____ , _____).

$64x - 56y = 0$
$-49x + 56y = -105$

Or $-64x + 56y = 0$
$49x - 56y = 105$.

350. Transform the system

$$8x - 7y = 0$$
$$7x - 8y = 15$$

to an equivalent system in which the terms containing y are the negatives of each other.

(−7,−8)

351. The solution of the system

$$64x - 56y = 0$$
$$-49x + 56y = -105$$

is (_____ , _____).

Remark. Is the idea here clear? What you are trying to do in solving a given system is to find an equivalent system in which the coefficients of one or the other of the variables are the negatives of each other. You can do this by multiplying both members of one or both of the equations by an appropriate constant. Once you have such a system, the addition of the corresponding members of the equations in the system will result in an equation in one variable only, which you can then solve with ease.

$(4, -4)$

232. Since 4 and -4 are the coordinates of the point of intersection of the graphs of $2x - y = 12$ and $2x + y = 4$, the solution of the system is (_____, _____).

$(3, 5)$

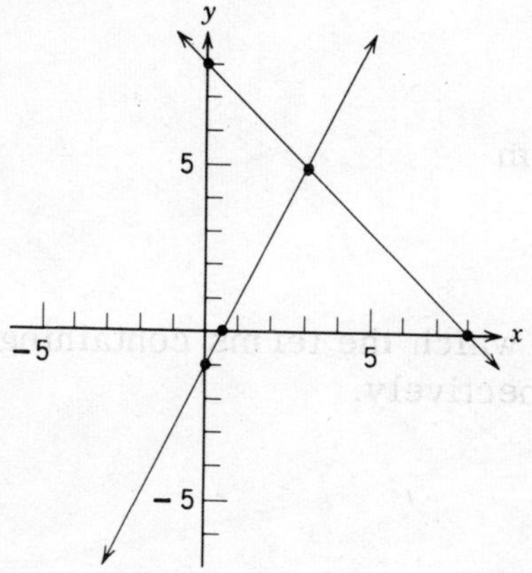

233. The graph of $2x - y = 1$ is shown. Graph $x + y = 8$ on the same coordinate system. The solution of the system is (_____, _____).

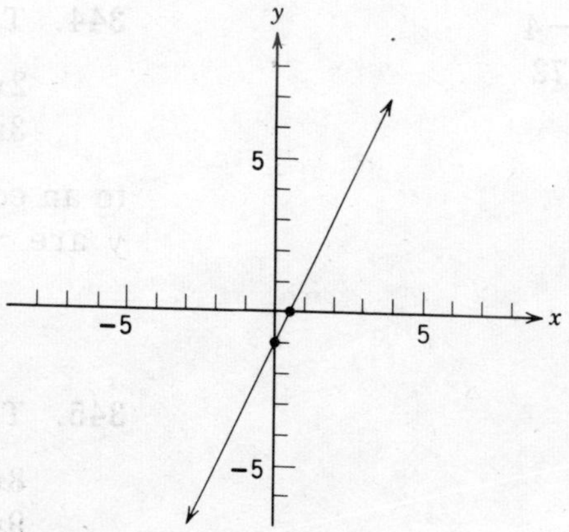

$(0, 3)$

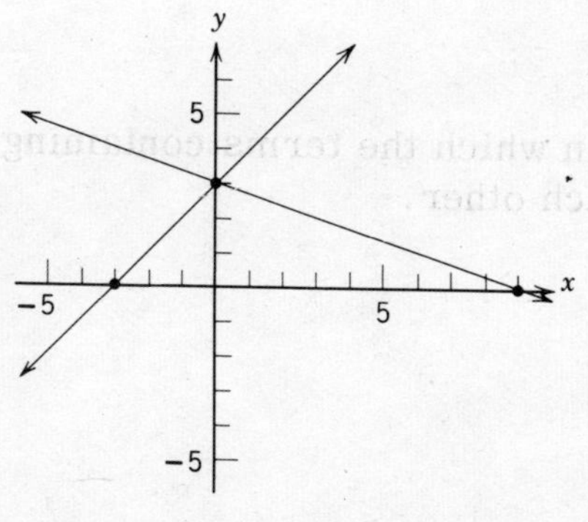

234. The graph of $x + 3y = 9$ is shown. Graph $y = 3 + x$ on the same coordinate system. The solution of the system is (_____, _____).

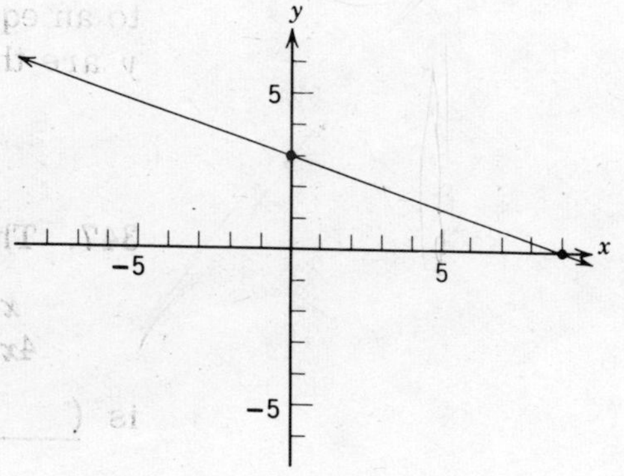

$10x - 6y = 18$
$6x + 6y = 30$

342. Transform the system

$$5x - 3y = 9$$
$$2x + 2y = 10$$

to an equivalent system in which the terms containing y are $-6y$ and $6y$, respectively.

(3, 2)

343. The solution of

$$10x - 6y = 18$$
$$6x + 6y = 30$$

is (____ , ____).

$8x - 12y = -4$
$9x + 12y = 72$

344. Transform the system

$$2x - 3y = -1$$
$$3x + 4y = 24$$

to an equivalent system in which the terms containing y are $-12y$ and $12y$, respectively.

(4, 3)

345. The solution of

$$8x - 12y = -4$$
$$9x + 12y = 72$$

is (____ , ____).

$x + 2y = 8$
$4x - 2y = 12$

346. Transform the system

$$x + 2y = 8$$
$$2x - y = 6$$

to an equivalent system in which the terms containing y are the negatives of each other.

(4, 2)

347. The solution of

$$x + 2y = 8$$
$$4x - 2y = 12$$

is (____ , ____).

$\left(-\frac{5}{2}, -3\right)$

Since you estimated the coordinates, your numbers may be slightly different from ours.

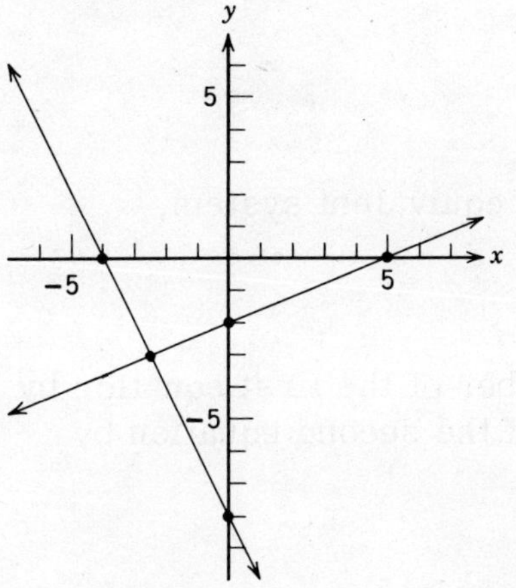

235. The graph of $2x - 5y = 10$ is shown. Graph $2x + y = -8$ on the same coordinate system. The solution of the system is $\left(\_\_\_\_ , \_\_\_\_ \right)$.

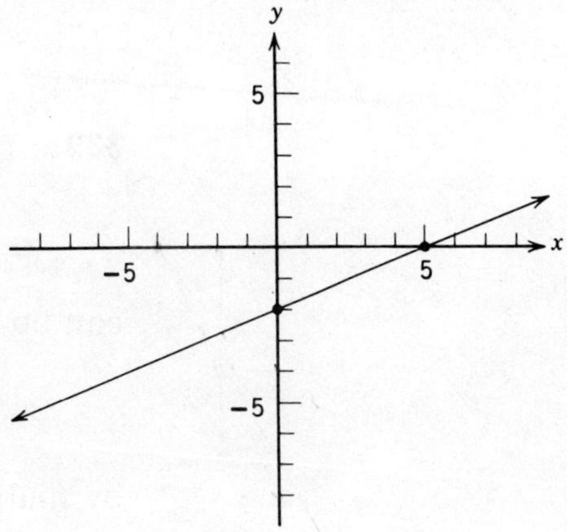

$(3, -1)$

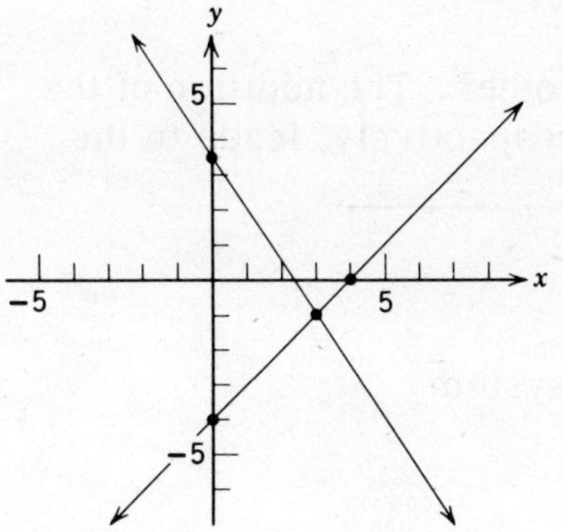

236. Graph $3x + 2y = 7$ and $x - y = 4$ on the same coordinate system. The solution of the system is $(\_\_\_\_ , \_\_\_\_)$.

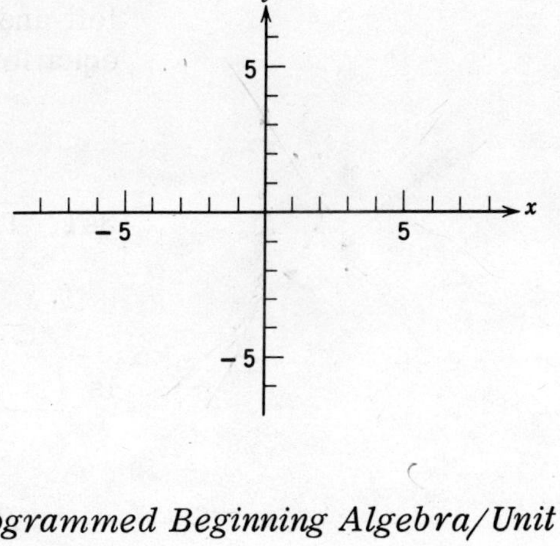

negatives

337. The coefficients of each term containing y in the system

$$8x - 12y = 4$$
$$9x + 12y = 30$$

are the____________________of each other.

(2, 1)

338. The solution of

$$8x - 12y = 4$$
$$9x + 12y = 30$$

is (____, ____).

3; 2

339. The system

$$3x - 2y = 3$$
$$-5x + 3y = -4$$

can be transformed to an equivalent system,

$$9x - 6y = 9$$
$$-10x + 6y = -8,$$

by multiplying each member of the first equation by _____and each member of the second equation by _____.

$-x = 1$

340. The coefficients of each term containing y in the system

$$9x - 6y = 9$$
$$-10x + 6y = -8$$

are the negatives of each other. The addition of the left and right members, respectively, leads to the equation in one variable, _____=_____.

(−1, −3)

341. The solution of the system

$$9x - 6y = 9$$
$$-10x + 6y = -8$$

is (____, ____).

(−4, −1)

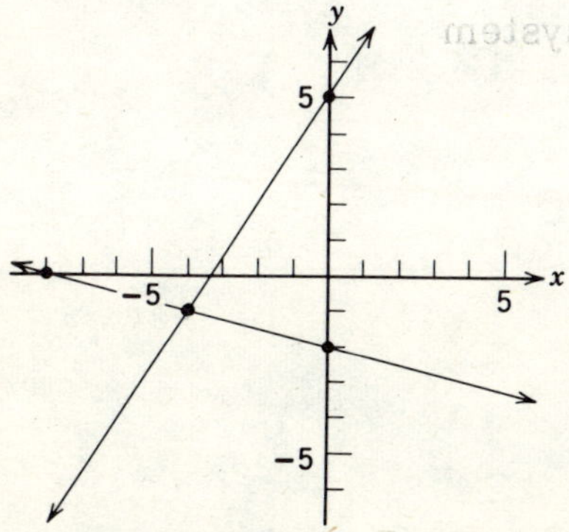

237. Graph $x + 4y = -8$ and $3x - 2y = -10$ on the same coordinate system. The solution of the system is (______, ______).

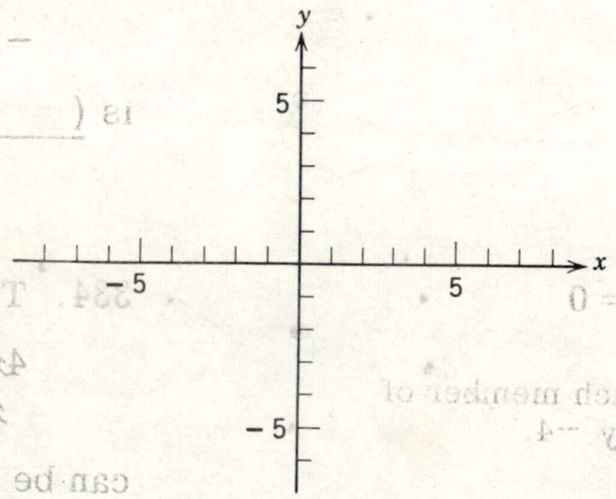

(−5, −2)

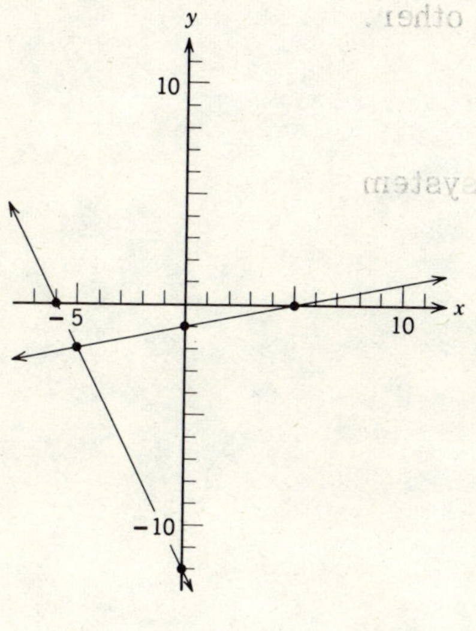

238. Find the solution of the system

$$x - 5y = 5$$
$$2x + y = -12$$

by graphical methods. The solution is (______, ______).

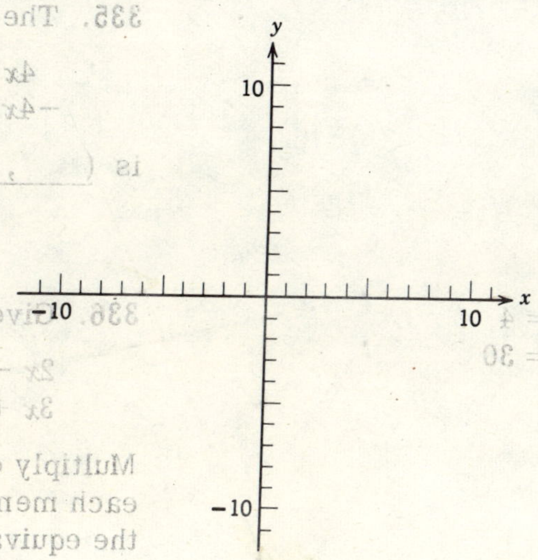

$-2x + 6y = 22$

Multiply each member of $-x + 3y = 11$ by 2.

332. The system

$$\begin{aligned} 2x + y &= 6 \\ -x + 3y &= 11 \end{aligned}$$

can be transformed to an equivalent system,

$$\begin{aligned} 2x + y &= 6 \\ \underline{\hspace{3em}} &= \underline{\hspace{2em}}, \end{aligned}$$

in which the coefficients of the terms containing x are the negatives of each other.

$(1, 4)$

333. The solution of the system

$$\begin{aligned} 2x + y &= 6 \\ -2x + 6y &= 22 \end{aligned}$$

is $(\underline{\quad}, \underline{\quad})$.

$-4x - 4y = 0$

Multiply each member of $x + y = 0$ by -4.

334. The system

$$\begin{aligned} 4x + 3y &= 1 \\ x + y &= 0 \end{aligned}$$

can be transformed to an equivalent system,

$$\begin{aligned} 4x + 3y &= 1 \\ \underline{\hspace{3em}} &= \underline{\hspace{2em}}, \end{aligned}$$

in which the coefficients of the terms containing x are the negatives of each other.

$(1, -1)$

335. The solution of the system

$$\begin{aligned} 4x + 3y &= 1 \\ -4x - 4y &= 0 \end{aligned}$$

is $(\underline{\quad}, \underline{\quad})$.

$8x - 12y = 4$
$9x + 12y = 30$

336. Given the system

$$\begin{aligned} 2x - 3y &= 1 \\ 3x + 4y &= 10. \end{aligned}$$

Multiply each member of the first equation by 4 and each member of the second equation by 3 and write the equivalent system obtained.

(5, 7)

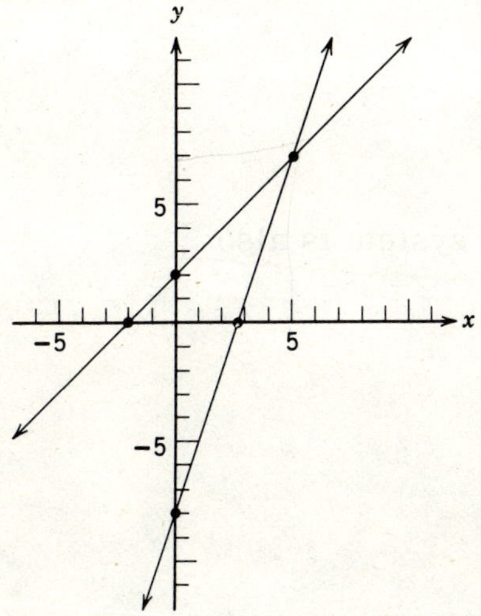

239. Find the solution of the system

$$x - y = -2$$
$$3x - y = 8$$

by graphical methods. The solution is (____ , ____).

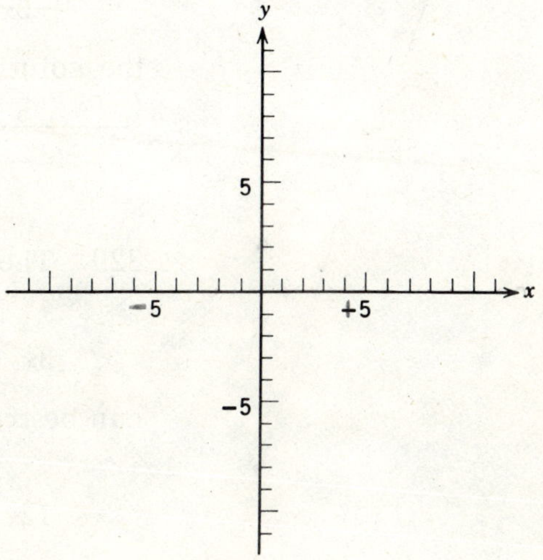

(4, 3)

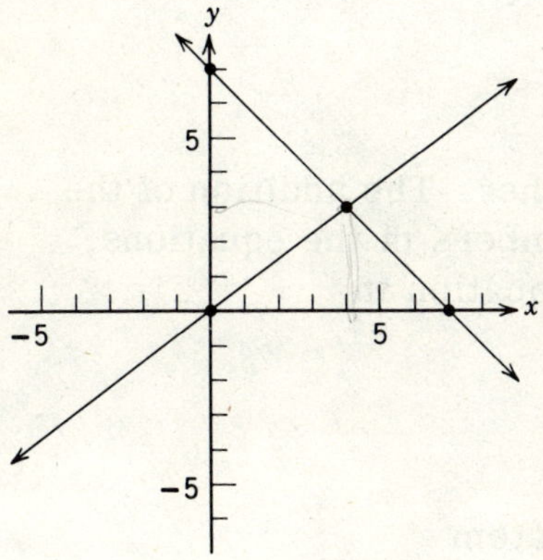

240. Solve

$$3x = 4y$$
$$x = 7 - y$$

by graphical methods. The solution is (____ , ____).

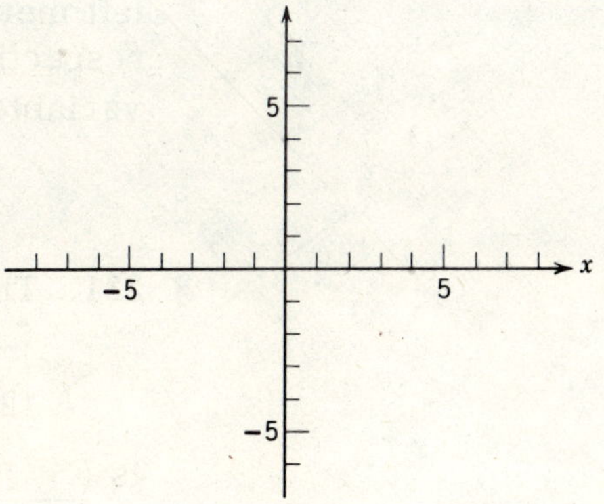

(10, −9)

328. Since the system

$$5x + 5y = 5$$
$$-5x - 6y = 4$$

is equivalent to the system

$$x + y = 1$$
$$-5x - 6y = 4,$$

the solution of the original system is also (____, ____).

4

329. The system

$$x - 4y = 4$$
$$3x + y = 12$$

can be transformed to the equivalent system

$$x - 4y = 4$$
$$12x + 4y = 48$$

by multiplying each member of $3x + 4y = 12$ by____ .

one ; $13x = 52$

330. The coefficients of the term containing y in the system

$$x - 4y = 4$$
$$12x + 4y = 48$$

are the negatives of each other. The addition of the left members and right members of the equations, respectively, leads to the equation in________ variable, __________.

(4, 0)

331. The solution of the system

$$x - 4y = 4$$
$$12x + 4y = 48$$

is (____, ____).

Remark. You can see now that an ability to graph a linear equation in two variables is a valuable talent to have. If you continue studying mathematics, you will find that there are many important uses for graphing. However, solving systems of linear equations by graphing is not the best way to go about it. In the first place, graphing takes time, and in the second place, you frequently have to estimate the solution. It would take a very sharp eye to be able to read a solution like $\left(\frac{231}{317}, -\frac{817}{913}\right)$ from a graph.

Fortunately, we have at our disposal other methods of solving a system of simultaneous equations, methods that are much more efficient than graphing. Let us look at one of them.

6 $5 + 1 = 6.$	**241.** The sum of the left members of the simultaneous equations $x - y = 1$ $x + y = 5$ is $x + x - y + y$ or $2x$; the sum of the right members is____.
$3x$; 9 $2x + x - 3y + 3y = 3x$; $4 + 5 = 9.$	**242.** The sum of the left members of the equations in the system $2x - 3y = 4$ $x + 3y = 5$ is______ ; the sum of the right members is____.
$3y$; 6 $x - x - 2y + 5y = 3y$; $-4 + 10 = 6.$	**243.** The sum of the left members of $x - 2y = -4$ $-x + 5y = 10$ is_______ ; the sum of the right members is____.

3

324. The system

$$x - 3y = 7$$
$$3x + y = 1$$

can be transformed to the equivalent system

$$x - 3y = 7$$
$$9x + 3y = 3,$$

where the terms containing the variable y are now the negatives of each other, by multiplying each member of the second equation by____.

negatives

325. The equations in the system

$$x + y = 1$$
$$-5x - 6y = 4$$

do not contain any terms that are the negatives of each other. However, if each member of the first equation is multiplied by 5, the terms containing the variable x in the resulting system will be the____________ of each other.

$$5x + 5y = 5$$
$$-5x - 6y = 4$$

326. If each member of the first equation in the system

$$x + y = 1$$
$$-5x - 6y = 4$$

is multiplied by 5, the result is a new system,

____________ = ____
____________ = ____,

which is equivalent to the given system.

(10, −9)

327. The solution of the system

$$5x + 5y = 5$$
$$-5x - 6y = 4$$

is (____, ____).

true

244. The statements $3 = 3$ and $4 = 4$ are true statements. Notice that if the left and right members of these statements are added, respectively, the result is $7 = 7$, which is also a ________statement.

addition

245. It is assumed that if a, b, c, and d are numbers, and if $a = b$ and $c = d$ are true statements, then the sums of their respective members is also a true statement; that is, $a + c = b + d$. This is called the addition axiom of equality. The assertion that if $x = 4$ and $y = 5$ are true statements, then $x + y = 9$ is true, is an application of the________________axiom of equality.

12

246. If $x = 4$ and $y = 8$, then the addition axiom of equality asserts that $x + y =$______.

$x + y = 10$

247. By the addition axiom of equality, if $x = 7$ and $y = 3$, then______________= 10.

$2x = 10$

248. For any values of the variables for which

$$x + y = 7$$
$$x - y = 3$$

are both true, the addition axiom of equality guarantees that $x + x + y - y = 10$ or__________= 10.

addition axiom

249. For any values of the variables for which

$$x + 2y = 5$$
$$-x + 3y = 10$$

are both true, the______________ ____________ of equality guarantees that $5y = 15$.

(3, 6)

319. Solve: $2x - y = 0$
$-6x + y = -12.$

(1, 1)

320. Solve: $2x + y = 3$
$y = 2 - x.$

(−4, 0)

321. Solve: $3x = -12 - 2y$
$x = -4 - 2y.$

(0, 1)

322. Solve: $x = 3y - 3$
$2y = 2 - x.$

Remark. You can now handle systems such as

$$x + 2y = 3$$
$$x - y = 0,$$

where there are identical terms in one variable in each equation. All you have to do is multiply one equation or the other by −1 and then apply the addition axiom. But, suppose there are no identical terms. For instance, the system

$$x - 3y = 7$$
$$3x + y = 1$$

cannot be solved by multiplying one of the equations by −1, because even after the multiplication is completed, you still cannot eliminate a variable by adding. What now? Well, we shall see.

3

323. The equation $3x + y = 1$ can be transformed to the equivalent equation $9x + 3y = 3$ by applying the multiplication axiom and multiplying each member by______.

solution

250. When the left and right members of simultaneous equations in two variables are added and their sums set equal to each other to obtain an equation in one variable, the solution of the equation in one variable is one component of the solution of the simultaneous equations. Adding the left and right members of

$$x - y = 8$$
$$x + y = 6$$

gives

$$2x = 14.$$

7 is a solution of $2x = 14$ and 7 is also the x-component of the ________________ of the simultaneous equations.

component

251. Setting the sum of the left members of the equations

$$x + y = 7$$
$$x - y = 3$$

equal to the sum of the right members leads to $2x = 10$. The solution of $2x = 10$ is 5, which is also one ________________ of the solution of the original system.

(5, 2)

x is 5 and y is 2.

252. Since $x = 5$ is one component of the solution of

$$x + y = 7$$
$$x - y = 3,$$

the solution is of the form $(5, y)$. The value for y can be found by replacing x by 5 in either of the equations and solving for y. If x is replaced by 5 in $x + y = 7$, y is 2. The complete solution is the ordered pair (____, ____).

$x - y = 3$

253. (5, 2) is a solution of

$$x + y = 7$$
$$x - y = 3$$

because (5, 2) is a solution of $x + y = 7$ and also a solution of ________ = ________.

$4x - 3y = 2$
$4x + y = -2$

314. Write the system

$$4x = 3y + 2$$
$$y + 2 = -4x$$

in standard form.

$4x - 3y = 2$
$-4x - y = 2$

315. The system

$$4x - 3y = 2$$
$$4x + y = -2$$

can be transformed to the system

______________ = ______
______________ = ______.

by multiplying each member of the second equation by -1.

one

316. The coefficients of the first term of the left member in the system

$$4x - 3y = 2$$
$$-4x - y = 2$$

are the negatives of each other. Adding the left members and right members, respectively, leads to the equation $-4y = 4$, which contains only_____ variable.

$\left(-\frac{1}{4}, -1\right)$

317. The solution of the system

$$4x - 3y = 2$$
$$-4x - y = 2$$

is $\left(\quad , \quad\right)$.

$2x - y = 0$
$-6x + y = -12$

Or $-2x + y = 0$
$6x - y = 12.$

318. Transform

$$y - 2x = 0$$
$$12 = -y + 6x$$

to an equivalent system in which each equation is in standard form and the coefficients of the term containing the variable y are the negatives of each other.

system

Or simultaneous system.

254. An application of the addition axiom of equality to the system

$$2x - 3y = -1$$
$$5x + 3y = 29$$

leads to $7x = 28$, whose solution is 4. 4 is also one component of the solution of the ______________ of equations.

(4, 3)

Substituting 4 for x in $2x - 3y = -1$ gives $2(4) - 3y = -1$ or $y = 3$.

255. Since 4 is one component (the value of x) of the solution of the system

$$2x - 3y = -1$$
$$5x + 3y = 29,$$

the solution is of the form $(4, y)$. The value for y can be found by replacing x by 4 in either of the equations and solving for y. The solution of the system is (____, ____).

$2x = 6$

256. The addition axiom of equality can be used to find one component of the solution of the system

$$x - y = 1$$
$$x + y = 5.$$

The addition axiom applies for any value of x and y for which both equations are true. Thus, applying the addition axiom to the system leads to the equation____________.

3

257. Since the x-component of any solution of

$$x - y = 1$$
$$x + y = 5$$

must satisfy $2x = 6$, the required value for x can be found by solving this latter equation. The solution of $2x = 6$ is____.

-1

310. The system

$$2x + y = 6$$
$$x + y = -3$$

can be transformed to the equivalent system

$$2x + y = 6$$
$$-x - y = 3$$

by multiplying each member of the second equation by ______.

$(9, -12)$

311. The solution of

$$2x + y = 6$$
$$-x - y = 3$$

is (______, ______).

multiplying; -1

Or, dividing; -1

312. The systems

$$2y = 3x + 4$$
$$x = 8 - 2y$$

and

$$-3x + 2y = 4$$
$$x + 2y = 8$$

are equivalent. The latter system can be transformed to

$$-3x + 2y = 4$$
$$-x - 2y = -8$$

by ______________ each member of the second equation by ______.

$\left(1, \frac{7}{2}\right)$

313. The solution of

$$-3x + 2y = 4$$
$$-x - 2y = -8$$

is (______, ______).

2; (3, 2)

258. Since any solution of

$$x - y = 1$$
$$x + y = 5$$

must be of the form $(3, y)$, by substituting 3 for x in either equation, the value for y is found to be____, and the solution of the system is (____, ____).

$x + y = 5$

259. The solution of the system

$$x - y = 1$$
$$x + y = 5$$

is (3, 2). That is, (3, 2) is a solution of $x - y = 1$ and is also a solution of________.

$3x = 12$

260. Given the system

$$x - y = -4$$
$$2x + y = 16$$

the sum of the left members is $3x$ and the sum of the right members is 12. Therefore, for the same values of the variables that are solutions of both equations, $3x =$ ______.

8; (4, 8)

261. Adding the left members and right members, respectively, of

$$x - y = -4$$
$$2x + y = 16,$$

yields $3x = 12$, whose solution is 4. Therefore, the solution of the system is of the form $(4, y)$. Substituting 4 for x in either equation, say $2x + y = 16$, gives $2(4) + y = 16$ and $y =$____. The solution of the system is (____, ____).

$(2, y)$

262. Given the system

$$x - 2y = -6$$
$$3x + 2y = 14.$$

Adding left members and right members, respectively, yields $4x = 8$. Therefore, the solution of the system must be of the form (____, y).

Remark. You have now learned to solve systems such as

$$3x + 2y = 7$$
$$x - 2y = -3,$$

where the coefficients of the terms in one of the variables are the negatives of each other. In this example, one equation contains $2y$ and the other contains $-2y$, and when the addition axiom is applied, these terms add to 0 so that the resulting equation, $4x = 4$, has only one variable. Unhappily, all systems of two equations in two variables do not share this property, as, for example,

$$2x + y = 7$$
$$x + y = 4.$$

However, there is a way to solve such systems, as we shall see over the next sequence of frames.

-1

306. The equation $-x + y = 4$ can be transformed to the equivalent equation $x - y = -4$ by multiplying or dividing each member by -1. $x - 2y = -3$ can be transformed to the equivalent equation $-x + 2y = 3$ by multiplying or dividing each member by_____.

equivalent; multiplying

Or dividing.

307. $4x - 3y = 6$ can be transformed to the ________________ equation $-4x + 3y = -6$ by ________________ each member by -1.

$-x - y = 0$

308. If each member of $x + y = 0$ is multiplied by -1, the resulting equivalent equation is____________.

$-x + 5y = -4$

309. If each member of $x - 5y = 4$ is multiplied by -1, the resulting equivalent equation is____________.

4; (2, 4)

263. The solution of the system

$$\begin{aligned} x - 2y &= -6 \\ 3x + 2y &= 14 \end{aligned}$$

is of the form $(2, y)$. Substituting 2 for x in either equation, y is found to be___ , and the solution of the system is (___ , ___).

$(x, 3)$

264. Given the system

$$\begin{aligned} 2x + 3y &= -3 \\ -2x + y &= 15. \end{aligned}$$

Adding left members and right members, respectively, yields $4y = 12$. Therefore, the solution of the system must be of the form $(x,$ ___).

-6; $(-6, 3)$

265. The solution of the system

$$\begin{aligned} 2x + 3y &= -3 \\ -2x + y &= 15 \end{aligned}$$

is of the form $(x, 3)$. Substituting 3 for y in either equation, x is found to be___ , and the solution of the system is (___ , ___).

$5y = 10$

266. The coefficients of the first terms in each left member of the system

$$\begin{aligned} 2x + 4y &= 6 \\ -2x + y &= 4 \end{aligned}$$

are the negatives of each other. Adding left members and right members, respectively, yields__________, which is an equation in one variable.

$(x, 2)$

267. The solution of $5y = 10$ is 2 and the solution of the system

$$\begin{aligned} 2x + 4y &= 6 \\ -2x + y &= 4 \end{aligned}$$

must be of the form $(x,$ ___).

(−3, −1)

300. The solution of

$$x - y = -2$$
$$3x + y = -10$$

is (_____, _____).

$x - 2y = -5$
$-x + y = 7$

301. Rewrite the system

$$x + 5 = 2y$$
$$y - 7 = x$$

in standard form.

(−9, −2)

302. Solve: $x - 2y = -5$
$-x + y = 7.$

(−7, −1)

Did you first rewrite the system in standard form?

303. Solve: $x = 4y - 3$
$y = x + 6.$

(3, 2)

304. Solve: $x + y = 5$
$x = 1 + y.$

(1, 1)

305. Solve: $y = 3 - 2x$
$2x = y + 1.$

-1; $(-1, 2)$

268. The solution of the system

$$2x + 4y = 6$$
$$-2x + y = 4$$

is of the form $(x, 2)$. Substituting 2 for y in either equation, x is found to be_____ , and the solution of the system is (______ , ______).

$(-1, 3)$

Adding the left members and the right members of the equations yields $7x = -7$.

269. The solution of the system

$$3x - y = -6$$
$$4x + y = -1$$

is (______ , ______).

$(1, 2)$

270. The solution of the system

$$5x - 3y = -1$$
$$3x + 3y = 9$$

is (______ , ______).

$(2, 2)$

271. Solve: $2x - 3y = -2$
$x + 3y = 8.$

$\left(-\frac{2}{5}, -2\right)$

272. Solve: $-5x + y = 0$
$5x - 6y = 10.$

$\left(-\frac{4}{3}, -2\right)$

273. Solve: $-3x + 7y = -10$
$3x + y = -2.$

$$\begin{aligned} x + y &= 5 \\ -x + y &= 7 \end{aligned}$$

Or $\begin{aligned} y + x &= 5 \\ y - x &= 7. \end{aligned}$

295. Rewrite the system

$$\begin{aligned} x + y &= 5 \\ y - x &= 7 \end{aligned}$$

in standard form.

$(-1, 6)$

296. To solve the system

$$\begin{aligned} x + y &= 5 \\ y - x &= 7, \end{aligned}$$

the system is first written in standard form,

$$\begin{aligned} x + y &= 5 \\ -x + y &= 7, \end{aligned}$$

and solved by applying the addition axiom of equality. The solution of the system is (_____, _____).

$$\begin{aligned} x + 3y &= 14 \\ 2x - 3y &= 4 \end{aligned}$$

297. Rewrite the system

$$\begin{aligned} x + 3y &= 14 \\ -3y &= 4 - 2x \end{aligned}$$

in standard form.

$\left(6, \frac{8}{3}\right)$

298. Solve the simultaneous equations

$$\begin{aligned} x + 3y &= 14 \\ 2x - 3y &= 4. \end{aligned}$$

The solution is (_____, _____).

$$\begin{aligned} x - y &= -2 \\ 3x + y &= -10 \end{aligned}$$

299. Rewrite the system

$$\begin{aligned} 2 - y &= -x \\ y &= -3x - 10 \end{aligned}$$

in standard form.

Remark. If the equations in a system do not happen to be written in a convenient form for solution, for example, the system

$$3x = 11 + 2y$$
$$-2y = 3x - 19,$$

it is necessary to find an equivalent system where the addition axiom can be applied directly. In this example, it would be convenient to use the equivalent system

$$3x - 2y = 11$$
$$-3x - 2y = -19.$$

In the following frames you will learn about equivalent systems and how to find them.

equivalent

274. Recall that equivalent equations are equations that have the same solution set. Thus, because $3x + x = y$ and $4x = y$ have the same solution set, they are__________________ equations.

equivalent

275. The application of the addition, multiplication, or division axioms of equality to an equation results in an equivalent equation. For example, if x is added to both members of $y = 3 - x$, the result is $x + y = 3$, which is__________________ to $y = 3 - x$.

solution

276. Equivalent equations have the same solution sets and the same graphs. Since $y = 3 - x$ and $x + y = 3$ are equivalent equations, they have the same ______________ sets and the same graphs.

equivalent

277. Since multiplying each member of $\frac{1}{3}x = y$ by 3 results in $x = 3y$, $\frac{1}{3}x = y$ and $x = 3y$ are __________________ equations.

solution; graphs

278. Because $\frac{1}{3}x = y$ and $x = 3y$ are equivalent equations, they have the same______________ sets and the same__________.

standard

290. A system of linear equations in which the terms containing the variables are written as the left member of each equation, and the constants as the right member, and in which like terms are vertically aligned, is said to be in standard form. Thus,

$$\begin{aligned} x - y &= 4 \\ 2x + y &= 8 \end{aligned}$$

is in ____________________ form.

standard form

291. The system of equations

$$\begin{aligned} x + 2y &= 4 \\ x - 3y &= -6 \end{aligned}$$

is in ____________________ ____________.

is

292. The system

$$\begin{aligned} 3x - 28y &= 4 \\ -8x + \quad y &= 0 \end{aligned}$$

(is/is not) in standard form.

is not

Like terms are not vertically aligned.

293. The system

$$\begin{aligned} x - \quad y &= 5 \\ y + 2x &= 6 \end{aligned}$$

(is/is not) in standard form.

$2x + y = 6$

294. Replace the second equation of the system

$$\begin{aligned} x - \quad y &= 5 \\ y + 2x &= 6 \end{aligned}$$

with an equivalent equation so that the system is in standard form.

$$\begin{aligned} x - \quad y &= 5 \\ &= \end{aligned}$$

equivalent

279. If both members of $2x + 2y = 4$ are divided by 2, the result is $x + y = 2$, which is ________________ to $2x + 2y = 4$.

y

280. Recall that by the commutative law of addition $y + x = x + y$. Therefore, $y + x = 6$ can be written as an equivalent equation $x + y = 6$, and $y - 2x = 3$ can be written as an equivalent equation $-2x +$ ______ $= 3$.

Just interchange $-2x$ and y in the equation $y - 2x = 3$.

$3x + 2y = 6$

281. Rewrite $2y + 3x = 6$ as an equivalent equation in which the term containing x is the first term in the left member, and the right member is 6.

$x - y = 0$

282. Rewrite $-y + x = 0$ as an equivalent equation in which the term containing x is the first term in the left member, and the right member is 0.

$4x$

283. The equation $y = 5 - 2x$ can be transformed to the equivalent equation $2x + y = 5$ by applying the addition axiom and adding $2x$ to each member. Similarly, the equation $y = 5 - 4x$ can be transformed to the equivalent equation $4x + y = 5$ by adding ______ to each member.

$-2y$

284. The equation $x = 6 + 2y$ can be written as an equivalent equation, $x - 2y = 6$, by adding ______ to each member.

$-3x + y = 4$

285. To transform $y = 3x + 4$ to an equivalent equation in which the term containing x is the first term in the left member, the addition axiom is applied. The result is the equation ____________ $= 4$.

$x + 2y = 4$

286. Transform $x = 4 - 2y$ to an equivalent equation in which the term containing x is the first term in the left member, and in which the right member is 4.

are

287. The equations $2x = y + 3$ and $2x - y = 3$ (are/are not) equivalent.

are

288. The equations $4y - 3 = -7x$ and $7x + 4y = 3$ (are/are not) equivalent.

Remark. Since two equivalent equations have identical solution sets, if any equation in a system is replaced with an equivalent equation, the resulting system is equivalent to (has the same solution set as) the original system. Because $x = 2y + 4$ is equivalent to $x - 2y = 4$, the system

$$x - 3y = 2$$
$$x = 2y + 4$$

is equivalent to the system

$$x - 3y = 2$$
$$x - 2y = 4.$$

$-x + 2y = -3$
$x - y = 4$

289. Because $2y = x - 3$ is equivalent to $-x + 2y = -3$, the system

$$2y = x - 3$$
$$x - y = 4$$

is equivalent to the system

$$\_\_\_\_ = -3$$
$$x - y = 4.$$